THE TRANSACTIONAL INTERPRETATION OF QUANTUM MECHANICS

A comprehensive exposition of the transactional interpretation of quantum mechanics (TI), this book sheds new light on longstanding problems in quantum theory and provides insight into the compatibility of TI with relativity. It breaks new ground in interpreting quantum theory, presenting a compelling new picture of quantum reality.

The book shows how TI can be used to solve the measurement problem of quantum mechanics, and to explain other puzzles, such as the origin of the "Born Rule" for the probabilities of measurement results. It addresses and resolves various objections and challenges to TI, such as Maudlin's inconsistency challenge. It explicitly extends TI into the relativistic domain, providing new insight into the basic compatibility of TI with relativity and the physical meaning of "virtual particles." This book is ideal for researchers and graduate students interested in the philosophy of physics and the interpretation of quantum mechanics.

RUTH E. KASTNER is a Research Associate and member of the Foundations of Physics group at the University of Maryland, College Park. She is the recipient of two National Science Foundation research awards for research in time symmetry issues and the transactional interpretation.

THE TRANSACTIONAL INTERPRETATION OF QUANTUM MECHANICS

The Reality of Possibility

RUTH E. KASTNER
University of Maryland, College Park

CAMBRIDGE UNIVERSITY PRESS
Cambridge, New York, Melbourne, Madrid, Cape Town,
Singapore, São Paulo, Delhi, Mexico City

Cambridge University Press
The Edinburgh Building, Cambridge CB2 8RU, UK

Published in the United States of America by Cambridge University Press, New York

www.cambridge.org
Information on this title: www.cambridge.org/9780521764155

First published 2013

Printed and bound in the United Kingdom by the MPG Books Group

A catalogue record for this publication is available from the British Library

Library of Congress Cataloging in Publication data
Kastner, Ruth E., 1955–
The transactional interpretation of quantum mechanics : the reality of possibility / Ruth E. Kastner.
p. cm.
Includes bibliographical references and index.
ISBN 978-0-521-76415-5 (Hardback)
1. Transactional interpretation (Quantum mechanics) I. Title.
QC174.17.T67K37 2012
530.12–dc23
2012013412

ISBN 978-0-521-76415-5 Hardback

Contents

Preface

This book came about as a result of my profound dissatisfaction with the existing "mainstream" interpretations of quantum theory and my conviction that the unusual mathematical structure of quantum theory indeed reflects something about physical reality, however subtle or hidden. In my early days as a physics graduate student, I was a "Bohmian"; however, I became dissatisfied with that interpretation for reasons discussed here and there throughout the book. It is my hope that, even if the reader does not come away convinced of the fruitfulness of the present approach, this presentation will serve as an invitation to further far-ranging and open discussion of the interpretational possibilities of quantum theory.

I have attempted to make much of the book accessible to the interested layperson with a mathematics and/or physics background, and to indicate where more technical sections can be omitted without losing track of the basic conceptual picture. For those in the field, I have endeavored to take into account as much as possible of the relevant literature and to use notes where a technical and/or esoteric point seems relevant. Chapters 5 and 6 are the most technical and may be omitted without losing track of the conceptual picture.

I am grateful to many colleagues, friends, and family members who gave generously of their time and energy to critically read drafts of various chapters, to offer comments, and to discuss material appearing herein. In particular, Professor John Cramer offered numerous suggestions for improvement of the manuscript, although we are not in agreement on all aspects of this proposal. His inclusion in the following list of acknowledgments therefore does not imply his endorsement of this formulation. Of course, final responsibility for the contents is mine alone.

My sincere thanks are owed to:

Stephen Brush
Leonardo Chiatti
John Cramer

Michael Devitt
Donatello Dolce
Avshalom Elitzur
Chris Fields
Michael Ibison
Joseph Kahr
Robert Klauber
Matt Leifer
James Malley
Louis Marchildon
David Miller
John Norton
Huw Price
Ross Rhodes
Troy Shinbrot
Michael Silberstein
Peter Evans
Peter Lewis
Steven Savitt
Eugene Solov'ev
Henry Stapp
William Unruh

Finally, I wish to thank my daughter, philosopher-artist Wendy Hagelgans, for valuable discussions concerning the nature of time and for drawing many of the images in this book, as well as friend and philosopher-artist Ty D'Avila for his insights and for allowing me to use his photo for two of the illustrations in Chapter 8. My other daughter, Janet, provided encouragement and inspiration by her example of perseverance in the face of challenge as she has pursued personal and career goals. My husband, Chuck, provided a sounding board as well as nonstop support and encouragement, as did my mother, Bernice Kastner. I would like to dedicate this book to my family, including the memory of my late father Sid Kastner, a physicist who was also fascinated by our elusive reality, seen and unseen.

1

Introduction: quantum peculiarities

1.1 Introduction

This book is an overview and further development of the transactional interpretation of quantum mechanics (TI), first proposed by John G. Cramer (1980, 1983, 1986, 1988). First, let's consider the question: why does quantum theory need an "interpretation"? The quick answer is that quantum theory is an abstract mathematical construct that happens to yield very accurate predictions of the behavior of large collections of identically prepared microscopic systems (such as atoms). But it is just that: a piece of mathematics (together with rules for its application). The interpretational task is to understand what the mathematics signifies physically; in other words, to be able to say *what* the theory's mathematical quantities represent in physical terms, and to understand why the theory works as well as it does. Yet quantum theory has been notoriously resistant to interpretation: most "common-sense" approaches to interpreting the theory result in paradoxes and riddles. This situation has resulted in a plethora of competing interpretations, some of which actually change the theory in either small or major ways. In contrast, TI (and its new version, "possibilist TI", or PTI) does not change the basic mathematical formalism; in that sense it can be considered a "pure" interpretation.

One rather popular approach is to suggest that quantum theory is not "complete" – that is, it lacks some component(s) which, if known, would resolve the paradoxes – and that is why it presents apparently insurmountable interpretational difficulties. Some current proposed interpretations, such as Bohm's theory, are essentially proposals for "completing" quantum theory by adding elements to it which (at least at first glance) seem to resolve some of the difficulties. (That particular approach will be discussed below, along with other "mainstream" interpretations.) In contrast to that view, this book explores the possibility that quantum mechanics *is* complete and that the challenge is to develop a new way of interpreting its message, even if that approach leads to a strange and completely unfamiliar metaphysical picture. Of course, strange metaphysical pictures in connection with quantum theory are nothing new: Bryce DeWitt's full-blown "many worlds interpretation" (MWI) is a prominent example that has entered the popular culture. However, I believe that TI does a better job by accounting for more of the quantum formalism, and that it resolves other issues facing MWI.

1.1.1 Quantum theory is about possibility

This work will explore the view that quantum theory is describing an unseen world of possibility which lies beneath, or beyond, our ordinary, experienced world of actuality. Such a step may, at first glance, seem far-fetched; perhaps even an act of extravagant metaphysical speculation. Yet there is a well-established body of philosophical literature supporting the view that it is meaningful and useful to talk about possible events, and even to regard them as real. For example, the pioneering work of David Lewis made a strong case for considering possible entities as real.[1] In Lewis' approach, those entities were "possible worlds": essentially different versions of our actual world of experience, varying over many (even infinite) alternative ways that "things might have been." My approach here is somewhat less extravagant:[2] I wish to view as physically real the possible quantum *events* that might be, or might have been, experienced. So, in this approach, *those possible events are real, but not actual; they exist, but not in spacetime*. The *actual* event is the one that is experienced and that can be said to exist as a component of spacetime. I thus dissent from the usual identification of "physical" with "actual": an entity can be physical without being actual. In more metaphorical language, we can think of the observable portion of reality (the actualized, spacetime-located portion) as the "tip of an iceberg," with the unobservable, unactualized, but still real, portion as the submerged part (see Figure 1.1).

Another way to understand the view presented here is in terms of Plato's original dichotomy between "appearance" and "reality." His famous allegory of the Cave proposed that we humans are like prisoners chained in a dark cave, watching and studying shadows flickering on a wall and thinking that those shadows are real objects. However, in reality (according to the allegory) the real objects are behind us, illuminated by a fire which casts their shadows on the wall upon which we gaze. The objects themselves are quite different from the appearances of their shadows (they are richer and more complex). While Plato thought of the "unseen" level of reality in terms of perfect forms, I propose that the reality giving rise to the "shadow"-objects that we see in our spacetime "cave" consists of the quantum

[1] Lewis' view is known as "modal realism" or "possibilist realism."

[2] So, for example, I will not need to defend the alleged existence of "that possible fat man in the doorway," from the "slum of possibles," a criticism of the modal realist approach by Quine ("On what there is," p. 15 in *From A Logical Point of View*, 1953).

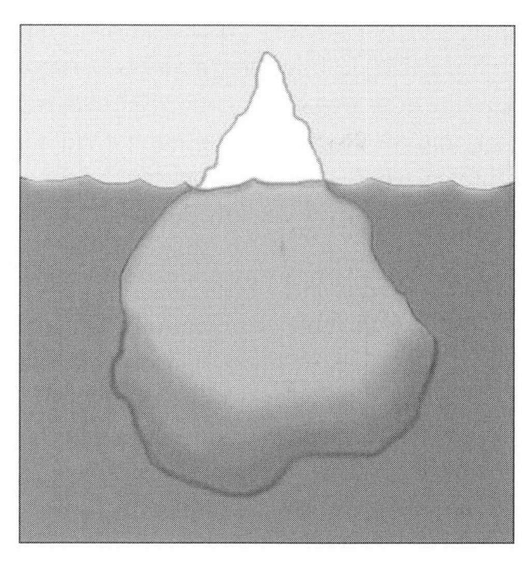

Figure 1.1 Possibilist TI: the observable world of spacetime events is the "tip of the iceberg" rooted in an unobservable manifold of possibilities transcending spacetime. These physical possibilities are what are described by quantum theory. (Drawing by Wendy Hagelgans.)

objects described by the mathematical forms of quantum theory. Because they are "too big," in a mathematical sense, to fit into spacetime (just as the objects casting the shadows are too big to fit on a wall in the cave, or the submerged portion of the iceberg cannot be seen above the water) – and thus cannot be fully "actualized" in the spacetime theater – we call them "possibilities." But they are *physically real* possibilities, in contrast to the way in which the term "possible" is usually used. Quantum possibilities are physically efficacious in that they *can* be actualized and thus can be experienced in the world of appearance (the empirical world).

This basic view will be further developed throughout the book. As a starting point, however, we need to take a broad overview of where we stand in the endeavor of interpreting the physical meaning of quantum theory. I begin with some notorious peculiarities of the theory.

1.2 Quantum peculiarities

1.2.1 Indeterminacy

The first peculiarity I will consider, *indeterminacy*, requires that I first discuss a key term used in quantum mechanics (QM), namely "*observable*." In ordinary classical physics, which describes macroscopic objects like baseballs and planets, it is easy to discuss the standard physical properties of objects (such as their position and momentum) as if those objects always possess determinate (i.e., well-defined, unambiguous) values. For example, in classical physics one can specify a baseball's position x and momentum p at any given time t. However, for reasons that will become clearer later on, in QM we cannot assume that the objects described by the theory – such as subatomic particles – always have such properties independently of interactions with, for example, a measuring device.[3] So, rather than talk about "properties," in QM we talk about "observables" – the things we can observe about a system based on measurements of it.

Now, applying the term "observable" to quantum objects under study seems to suggest that their nature is dependent on observation, where the latter is usually understood in an anthropocentric sense, as in observation by a conscious observer. The technical philosophical term for the idea that the nature of objects depends on how (or whether) they are perceived is "antirealism." The term "realism" denotes the opposite view: that objects have whatever properties they have independent of how (or whether) they are perceived: i.e., that the real status or nature of objects does not depend on their perception.

The antirealist flavor of the term "observable" in quantum theory has led researchers of a realist persuasion – a prominent example being John S. Bell – to be highly critical of the term. Indeed, Bell rejected the term "observable," and proposed instead a realist alternative, "beable." Bell intended "beable" to denote real properties of quantum objects that are independent of whether or not they are measured (one example being Bohmian particle positions; see Section 1.3.3). The interpretation presented in this book does not make use of "beables," although it shares Bell's realist motivation: quantum theory – by virtue of its impeccable ability to make accurate predictions about the phenomena we can observe – is telling us something about reality, and it is our job to discover what that might be, no matter how strange it may seem.[4]

I will address in more detail the issue of how to understand what an "observable" is in the context of the transactional interpretation in later chapters. For now, I simply deal with the perplexing issue of indeterminacy concerning the values of observables, as in the usual account of QM.

Heisenberg's famous "uncertainty principle" (also called the "indeterminacy principle") states that, for a given quantum system, one cannot simultaneously

[3] The apparent "cut" between macroscopic (e.g., a measuring device) and microscopic (e.g., a subatomic particle) realms has been one of the central puzzles of quantum theory. We will see (in Chapter 3) that under the transactional interpretation, this problem is solved; the demarcation between quantum and classical realms need not be arbitrary (or based on a subjectivist appeal to an observing "consciousness").

[4] The realist accounts for the success of a theory in a simply way: it describes something about reality. Antirealist and pragmatic approaches such as "instrumentalism" – that theories are just instruments to predict phenomena – can provide no explanation for why the successful theory works better than a competing theory. A typical account in support of such approaches would say that the demand for an explanation for why the theory works simply need not be met. I view this as an evasion of a perfectly legitimate, indeed crucial, question.

determine physical values for pairs of incompatible observables. "Incompatible" means that the observables cannot be simultaneously measured, and that the results one obtains depend on the order in which they are measured. Elementary particle theorist Joseph Sucher has a colorful way of describing this property. He observes that there is a big difference between the following two processes: (1) opening a window and sticking your head out, and (2) sticking your head out and then opening the window.[5]

Mathematically, the *operators* (i.e., the formal objects representing observables) corresponding to incompatible observables do not commute:[6] i.e., the results of multiplying such operators together depend on their order. Concrete examples are position, whose mathematical operator is denoted X (technically, the operator is really multiplication by position x), and momentum, whose operator is denoted P.[7] The fact that X and P do not commute can be symbolized by the statement

$$\mathrm{XP} \neq \mathrm{PX}$$

Thus, quantum mechanical observables are not ordinary numbers that can be multiplied in any order with the same result; instead, you must be careful about the order in which they are multiplied.

It is important to understand that the uncertainty principle is something much stronger (and *stranger*) than the statement that we just can't physically measure, say, both position and momentum because measuring one property disturbs the other one and changes it. Rather, in a fundamental sense, the quantum object *does not have* a determinate (well-defined) value of momentum when its position is detected, and vice versa. This aspect of quantum theory is built into the very mathematical structure of the theory, which says in precise logical terms that there simply is no yes/no answer to a question about the value of a quantum object's position when you are measuring its momentum. That is, the question "Is the particle at position x?" generally has no yes or no answer in quantum theory in the context of a momentum measurement. This is the puzzle of quantum indeterminacy: quantum objects seem not to have precise properties independent of specific measurements which measure those specific properties.[8]

A particularly striking example of indeterminacy on the part of quantum objects is exhibited in the famous two-slit experiment (Figure 1.2). This experiment is often discussed in conjunction with the idea of "wave/particle duality," which is a

[5] Comment by Professor Joseph Sucher in a 1993 UMCP quantum mechanics course.

[6] "Commute" literally means "go back and forth"; so that the standard commuting property is expressed by noting that for two ordinary numbers a and b, $ab = ba$.

[7] The mathematical form of P (in one spatial dimension) is given by $\mathrm{P} = \frac{\hbar}{i}\frac{d}{dx}$.

[8] Or properties belonging to a compatible observable (whose operator commutes with the one being measured). Bohmians dissent from this characterization of the theory; this will be discussed below.

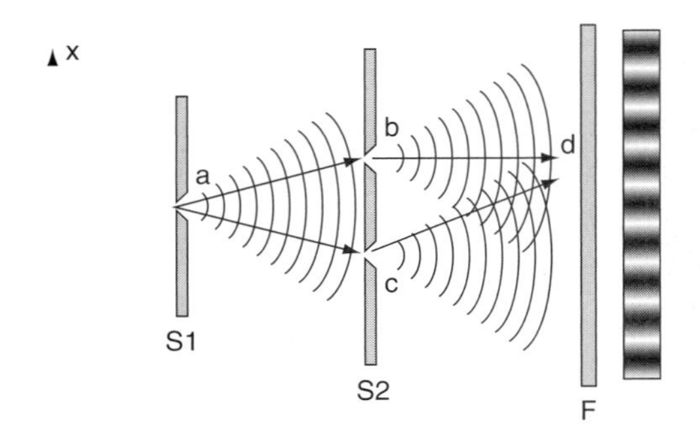

Figure 1.2 The double-slit experiment.
Source: http://en.wikipedia.org/wiki/File:Doubleslit.svg

manifestation of indeterminacy. (The experiment and its implications for quantum objects are discussed in the Feynman Lectures, Vol. 3, chapter 1 (Feynman *et al.*, 1964); I revisit this example in more detail in Chapter 3.)

If we shine a beam of ordinary light through two narrow slits, we will see an interference pattern (see Figure 1.2). This is because light behaves (under some circumstances) like a wave, and waves exhibit interference effects. A key revelation of quantum theory is that material objects (that is, objects with non-zero rest mass, in contrast to light) also exhibit wave aspects. So one can do the two-slit experiment with quantum particles as well, such as electrons, and obtain interference. Such an experiment was first performed by Davisson and Germer in 1928, and was an important confirmation of Louis de Broglie's hypothesis that matter also possesses wavelike properties.[9]

The puzzling thing about the two-slit experiment performed with material particles is that it is hard to understand what is "interfering": our classical common sense tells us that electrons and other material particles are like tiny billiard balls that follow a clear trajectory through such an apparatus. In that picture, the electron must go through one slit or the other. But if one assumes that this is the case and calculates the expected pattern, the result will *not* be an interference pattern. Moreover, if one tries to "catch it in the act" by observing which slit the electron went through, this procedure will ruin the interference pattern. It turns out that interference is seen only when the electron is left undisturbed, so that in some sense it "goes through both slits." Note that the interference pattern can be slowly built up dot by dot, with only one particle in the apparatus at a time (see Figure 1.3). Each of those dots represents an entity that is somehow "interfering with itself" and represents a particle whose

[9] Davisson, C. J. (1928) "Are electrons waves?," *Franklin Institute Journal* **205**, 597.

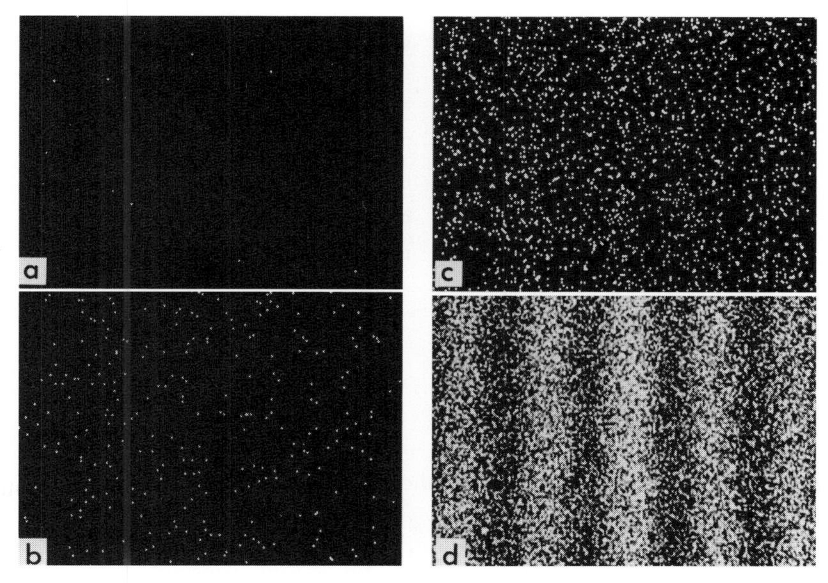

Figure 1.3 Results of a double-slit experiment performed by Dr Tonomura showing the build-up of an interference pattern of single electrons. Numbers of electrons are 11 (a), 200 (b), 6000 (c), 40 000 (d), 140 000 (e).
Source: Reprinted courtesy of Dr Akira Tonomura, Hitachi Ltd, Japan

position is indeterminate – it does not have a well-defined trajectory, in contrast to our classical expectations.[10]

1.2.2 Non-locality

The puzzle of non-locality arises in the context of composite quantum systems: that is, systems that are composed of two or more quantum objects. The prototypical example of non-locality is the famous Einstein–Podolsky–Rosen (EPR) paradox, first presented in a 1935 paper written by these three authors (Einstein *et al.*, 1935). The paper, entitled "Can quantum-mechanical description of reality be considered complete?," attempted to demonstrate that QM could not be a complete description of reality because it failed to provide values for physical quantities that the authors assumed must exist.

Here is the EPR thought-experiment in a simplified form due to David Bohm, in terms of spin-1/2 particles such as electrons. Spin-1/2 particles have the property

[10] One of the interpretations I will discuss, the Bohmian theory, does offer an account in which particles follow determinate trajectories. The price for this is a kind of non-locality that may be difficult to reconcile with relativity, in contrast to TI.

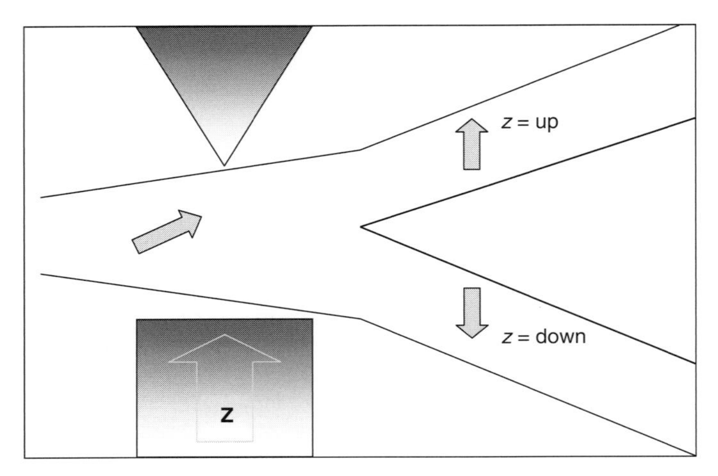

Figure 1.4 Spin "up" or "down" along the z direction in a SG measurement.

that, when subject to a non-uniform magnetic field along a certain spatial direction z, they can either align with the field (which is termed "up" for short) or against the field (termed "down") (see Figure 1.4).

I designate the corresponding quantum states as "$|z\ \text{up}\rangle$" and "$|z\ \text{down}\rangle$," respectively. The notation used here is the bracket notation invented by Dirac, and the part pointing to the right is the "$|\text{ket}\rangle$." We can also have a part pointing to the left, "$\langle\text{brac}|$." (Since one is often working with the inner product form $\langle\text{brac}|\text{ket}\rangle$, the name is an apt one.) We could measure the spin and find a corresponding result of either "up" or "down" along any direction we wish, by orienting the field along a different spatial direction, say x. The states we could then measure would be called "$|x\ \text{up}\rangle$" or "$|x\ \text{down}\rangle$," and similarly for any other chosen direction.

We also need to start with a composite system of two electrons in a special type of state, called an "entangled state." This is a state of the composite system that cannot be expressed as a simple, factorizable combination (technically a "product state") of the two electrons in determinate spin states, such as "$|x\ \text{up}\rangle|x\ \text{down}\rangle$."

If we denote the special state by $|\text{S}\rangle$, it looks like

$$|\text{S}\rangle = \frac{1}{\sqrt{2}}\left[|\text{up}\rangle|\text{down}\rangle - |\text{down}\rangle|\text{up}\rangle\right] \tag{1.1}$$

where no directions have been specified, since this state is not committed to any specific direction. That is, you could put in any direction you wish (provided you use the same "up/down – down/up" form); the state is mathematically equivalent for all directions.

Now, suppose you create this composite system at the 50-yard line of a football field and direct each of the component particles in opposite directions, say to two observers "Alice" and "Bob" in the touchdown zones at opposite ends of the field. Alice and Bob are each equipped with a measuring apparatus that can generate a local non-uniform magnetic field along any direction of their choice (as illustrated in Figure 1.4). Suppose Alice chooses to measure her electron's spin in the z direction. Then quantum mechanics dictates that the spin of Bob's particle, if measured along z as well, must always be found in the opposite orientation from Alice's: if Alice's electron turns out to be $|z\text{ up}\rangle$, then Bob's electron must be $|z\text{ down}\rangle$, and vice versa. The same holds for any direction chosen by Alice. Thus it seems as though Bob's particle must somehow "know" about the measurement performed by Alice and her result, even though it may be too far away for a light signal to reach in time to communicate the required outcome seen by Bob. This apparent transfer of information at a speed greater than the speed of light ($c = 3\times10^8$ m/s) is termed a "non-local influence," and this apparent conflict of quantum theory with the prohibition of signals faster than light is termed "non-locality."[11]

Einstein termed this phenomenon "spooky action at a distance" and used it to argue that there had to be something "incomplete" about quantum theory, since in his words, "no reasonable theory of reality should be expected to permit this."[12] However, it turns out that we are indeed stuck with quantum mechanics as our best theory of (micro)-reality despite the fact that it does, and must, permit this, as Bell's Theorem (1964) demonstrated. Bell famously showed that no theory that attributes local "elements of reality" of the kind presumed by Einstein to exist can reproduce the well-corroborated predictions of quantum theory; specifically, the strong correlations inherent in the EPR experiment. *Quantum mechanics is decisively non-local*: the components of composite systems described by certain kinds of quantum states (such as the state (1.1)) seem to be in direct, instantaneous communication with one another, regardless of how far they may be spatially separated.[13] The interpretational challenge presented by the EPR thought-experiment combined with Bell's Theorem is that a well-corroborated theory seems to show that reality *is* indeed

[11] I say "apparent conflict" here because it is a very subtle question as to what constitutes a genuine violation of, or conflict with, relativity. It is suggested in Section 6.4.2 that PTI can provide "peaceful coexistence" of QM with relativity, as envisioned by Shimony (2009).

[12] I am glossing over some subtleties here concerning Einstein's objection. A more detailed account of the EPR paper would note that Einstein's objection was in terms of "elements of reality" concerning the presumably determinate physical spin attributes of either electron and the fact that their quantum states seemed not to be able to specify these. As noted in the subsequent discussion, Bell's Theorem of 1964 showed that there can be no such "elements of reality."

[13] I should note that a small minority of researchers dissent from this characterization. A way out of the conclusion that quantum theory is necessarily non-local is to dispute the way "elements of reality" are defined. See, for example, Willem M. de Muynck's discussion at http://www.phys.tue.nl/ktn/Wim/qm4.htm!thermo_analogy. I am skeptical of this approach because it must introduce what appears to be an ad hoc further level of statistical randomness, beyond that of the standard theory, whose sole purpose is to enforce locality.

"unreasonable," in that it allows influences at apparently infinite (or at least much faster than light) speeds, despite the fact that relativity seems to say that such things are forbidden.

1.2.3 The measurement problem

If indeterminacy and non-locality seem to violate common sense, one should prepare for further violations of common sense in what follows. The measurement problem is probably the most perplexing feature of quantum theory. There is a vast literature on this topic, testifying to the numerous and sustained attempts to solve this problem. Erwin Schrödinger's famous "cat" example, which I will describe below, was intended by him to be a dramatic illustration of the measurement problem (Schrödinger, 1935).

The measurement problem is related to quantum indeterminacy in the following way. Our everyday experiences of always-determinate (clearly defined, non-fuzzy) properties of objects seems inconsistent with the mathematical structure of the theory, which dictates that sometimes such properties are *not* determinate. The latter cases are expressed as superpositions of two or more clearly defined states. For example, a state of indeterminate position, let's call it "$|?\rangle$," could be represented in terms of two possible positions x and y by

$$|?\rangle = a|x\rangle + b|y\rangle \tag{1.2}$$

where a and b are two complex numbers called "amplitudes." A quantum system could undergo some preparation leaving it in this state. If we wanted to find out where the system was, we could measure its position and, according to the orthodox way of thinking about quantum theory, its state would "collapse" into either position x or position y.[14] The idea that a system's state must "collapse" in this way upon measurement is called the "collapse postulate" (see Section 1.3.4) and is a matter of some controversy. Schrödinger's cat makes the controversy evident. I now turn to this famous thought-experiment.

Here is a brief description of the idea (with apologies to cat lovers). A cat is placed in a box containing an unstable radioactive atom which has a 50% chance of decaying (emitting a subatomic particle) within an hour. A Geiger counter, which detects such particles, is placed next to the atom. If a click is registered indicating

[14] The probability of ending up in x would be a^*a and in y would be b^*b. This prescription for taking the absolute square of the amplitude of the term to get the probability of the corresponding result is called the "Born Rule" after Max Born who first proposed it. Amplitudes are therefore also referred to as "probability amplitudes." There is no way to predict which outcome will result in any individual case. TI provides a concrete, physical (as opposed to statistical or decision-theoretic) basis for the Born Rule.

that the atom has decayed, a hammer is released which smashes a vial of poison gas, killing the cat. Otherwise, nothing happens to the cat. With this setup, we place all ingredients in the box, close it, and wait one hour.

The atom's state is usually written as a superposition of "undecayed" and "decayed," analogous to state (1.2):

$$|\text{atom}\rangle = \frac{1}{\sqrt{2}}\left[|\text{undecayed}\rangle + |\text{decayed}\rangle\right] \tag{1.3}$$

Prior to our opening the box, since no measurement has been performed to "collapse" this superposition, we are (so the usual story goes[15]) obligated to include the cat's state in the superposition as follows:

$$|\text{atom} + \text{cat}\rangle = \frac{1}{\sqrt{2}}\left[|\text{undecayed}\rangle|\text{alive}\rangle + |\text{decayed}\rangle|\text{dead}\rangle\right] \tag{1.4}$$

This superposition is assumed to persist because no "measurement" has occurred which would "collapse" the state into either alternative. So we appear to end up with a cat in a superposition of "alive" and "dead" until we open the box and see which it is, upon which the state of the entire system (atom + Geiger counter + hammer + gas vial + cat) "collapses" into a determinate result. Schrödinger's example famously illustrated his exasperation with the idea that something macroscopic like a cat seems to be forced into a bizarre superposition of alive and dead by the dictates of quantum theory, and that it is only when somebody "looks" at it that the superposed system is found to have collapsed, even though this mysterious "collapse" is never observed nor (apparently) is there any physical mechanism for it. This is the core of the measurement problem.

In less colorful language, the measurement problem consists in the fact that, given an initial quantum state for a system, quantum theory does not tell us *why* or *how* we only get one specific outcome when we perform a measurement on that system. On the contrary, the quantum formalism seems to tell us about several possible outcomes, each with a particular weight. So, for example, I could prepare a quantum system in some arbitrary state X, perform a measurement on it, and the theory would

[15] TI does not have to tell the story this way; in TI one does not need to characterize the system by equation (1.4). This fact, a major reason to choose TI over its competitors, is discussed in Chapters 3 and 4. A key component of the puzzle raised by Schrödinger's cat is that it is not at all obvious that a macroscopic object like a cat should be describable by a quantum state as in (1.4) (indeed, I argue that it is *not*). While many current approaches recognize this issue and try to address it, I believe that TI's approach is the only non-circular and unambiguous one, especially in view of Fields' criticism of the decoherence arguments (see Section 1.3.1) which underlie those competing approaches.

tell me that it might be A, or B, or C, but it will not tell me *which* result actually occurs, nor does it provide any reason for *why* only one of these is actually observed.

So there seems to be a very big and mysterious gap between what the theory appears to be saying (at least according to the usual understanding of it) and what our experience tells us in everyday life. We are technically sophisticated enough to create and manipulate microscopic quantum systems in the laboratory, to the extent that we can identify them with a particular quantum state (such as X above, for example). We can then put these prepared systems through various experimental situations intended to measure their properties. But, in general, for any of those measurements, the theory just gives us a weighted list of possible outcomes. And obviously, in the laboratory, we see only *one* particular outcome.

Now, the theory is still firmly corroborated in the sense that the weights give extremely accurate predictions for the *probabilities* of those outcomes when we perform the same kind of measurement on a large number of identically prepared systems (technically known as an *ensemble*). But the measurement problem consists in the fact that any individual system is still described by the theory, yet the theory does not specify what that individual system's outcome will actually be, or even why it has only one.

It should be emphasized that this situation is completely different from what classical physics tells us. For example, consider a coin flip. A coin is a macroscopic object that is well described by classical physics. If we knew everything about all the (classical) forces acting on the coin, and all the relevant details of the coin itself, we could in principle calculate the result of any particular coin flip. That is, we could predict with 100% certainty (or at least within experimental error) whether it would land heads or tails. But when it comes to the microscopic objects described by quantum theory, even if we start with precise knowledge of their initial states, in general the theory does not allow us to predict *any* given outcome with 100% certainty.[16] The situation is made even more perplexing by the fact that classical physics and quantum physics must be describing the same world, so they must be compatible in the limit of macroscopic objects (that is, when the sizes of our systems become much larger than subatomic particles like electrons and neutrons). This means that macroscopic objects must also be describable (in that same limit) by quantum theory. This consideration raises the important question of: "exactly *what* is a 'macroscopic object' anyway, and how is it different from the objects (like electrons) that can *only* be described by quantum theory?" The quick answer, under TI, is that macroscopic objects are phenomena resulting from actualized transactions, whereas quantum objects are not. I explore this in detail in Chapter 7.

[16] The exception, of course, is that measurements of observables commuting with the preparation observable result in determinate outcomes.

Typical prevailing interpretations even encounter difficulty in specifying exactly what counts as a measurement, and consider that question to be a component of the measurement problem. For example, discussions of the Schrödinger's cat paradox have dealt not only with the bizarre notion of a cat seemingly in a quantum superposition, but also with the conundrum of *when or how* measurement of the system can be considered truly finished. That is, does the observer who opens the box and looks at the cat also enter into a superposition? At what point does this superposition really "collapse" into a determinate (unambiguous) result? An example of this statement of the problem in the literature is provided by Clifton and Monton (1999):

> Unfortunately, the standard dynamics [and the standard way of interpreting] quantum states together give rise to the measurement problem; they force the conclusion that a cat can be neither alive nor dead, and, worse, that a competent observer who looks upon such a cat will neither believe that the cat is alive nor believe it to be dead. The standard way out of the measurement problem is to . . . temporarily suspend the standard dynamics by invoking the collapse postulate. According to the postulate, the state vector $|\psi(t)\rangle$, representing a composite interacting "measured" and "measuring" system, stochastically [randomly] collapses, at some time t' during their interaction . . . The trouble is that this is not a way out unless one can specify the physical conditions necessary and sufficient for a measurement interaction to occur; for surely "measurement" is too ambiguous a concept to be taken as primitive in a fundamental physical theory. (p. 698)

We will see in Chapters 3 and 4 that TI has a very effective "way out" of this conundrum, including the puzzle of defining what constitutes a "measurement." But for now, I just note that in view of the highly perplexing and seemingly intractable nature of the measurement problem, probably the most fervently sought-after feature of an interpretation of quantum theory is that it should provide a solution to this problem. A "solution to the measurement problem" is usually understood to be an explanation for how quantum theory's list of weighted outcomes (rather than a single determinate outcome) can be reconciled with our experience.

Peter Lewis (2007) has suggested that there are traditionally two basic conditions that need to be met by such an explanation:

Condition (1): the explanation must be consistent with other well-established physical theories, in particular the theory of relativity.

Condition (2): it must be consistent with basic philosophical commitments concerning reality.

Now, condition (1) is straightforward enough – although notoriously difficult to satisfy in prevailing interpretations – and part of this work will be dedicated to fulfilling that condition. However, condition (2) is where, in my view, the real conceptual challenge lies. The main thesis of this work is the claim that the apparently intractable nature of the measurement problem can be traced to the generally unrecognized need to substantially alter one or more of our "basic philosophical commitments concerning reality" in order to properly understand what the theory might be telling us. Before I address in detail what I think needs to be altered among those basic philosophical commitments, I briefly review some of the better-known interpretational approaches to "solving the measurement problem."

1.3 Prevailing interpretations of QM

1.3.1 Decoherence approaches

"Decoherence" refers to the way in which interference effects (like what we see in a two-slit experiment, Figures 1.2 and 1.3) are lost as a given quantum system interacts with its environment. Roughly speaking, decoherence amounts to the loss of the ability of the system to "interfere with itself" as the electron does in the two-slit experiment. This basic idea – that a quantum system suffers decoherence when it interacts with its environment – has been developed to a high technical degree in recent decades. In effect this research has shown that in most cases, quantum systems cannot maintain coherence, and its attendant interference effects, in processes which amplify such systems to the observable level of ordinary experience. In general, this approach to the classical level is described by a greatly increasing number of "degrees of freedom" of the system(s) under study.[17] So, decoherence shows that systems with many degrees of freedom – macroscopic systems – do not exhibit observable interference. In addition, the decoherence approach seems to provide a way to specify a determinate "pointer observable" for the apparatus used to measure a given system once the interactions of the system, apparatus, and environment are all taken into account. This apparent emergence via the decoherence process of a clearly defined, macroscopic "pointer observable" for a given measurement interaction is sometimes referred to as "quantum Darwinism," since the process seems analogous to an evolutionary process.

Many researchers have taken this as at least a partial solution to the measurement problem in that it is taken to explain why we don't see interference effects happening all around us even though matter is known to have wavelike properties. It appears to explain, for example, why Schrödinger's cat need not be thought of as exhibiting an interference pattern (which is something of a relief). But decoherence

[17] "Degrees of freedom" basically means "ways in which an object can move." A system of one particle (neglecting spin) can move in a spatial sense (in three possible directions), so it has three degrees of freedom. A system of three particles has nine degrees of freedom, and so on. If one assumes that the particles have spin, then additional, rotational degrees of freedom are in play.

alone does not explain why the cat is clearly *either* alive *or* dead (and not in some superposition) at the end of the experiment. The reason for this is somewhat technical, and amounts to the fact that we can still have quantum superpositions without interference. Such superpositions cannot be thought of as representing only an epistemic uncertainty (uncertainty based only on lack of knowledge about something that really is determinate). In order to regain the classical world of ordinary experience, we need to be able to say that our uncertainty about the status of an object is entirely epistemic – it is just our ignorance about the object's properties – and not based on an indeterminacy inherent in the object itself. Decoherence fails to provide this.

Here is a crude way to understand the distinction between merely epistemic uncertainty and quantum (objective) indeterminacy. Suppose I put 10 marbles in an opaque box; 3 red and 7 green, and then close the box. I could represent my uncertainty about the color of any particular marble I might reach in and grab by a statistical "mixture" of 30% red and 70% green. My uncertainty about those marbles is entirely contained in my ignorance about which one I will happen to touch first. There is nothing "uncertain" about the marbles themselves. Not so with a quantum system prepared in a state, say,

$$|\Psi>= a|\text{red}> + b|\text{green}>$$

We may be able to eliminate all interference effects from phenomena based on this object's interactions with macroscopic objects, but we have not eliminated the quantum superposition based on its state. In some sense, the state describes an *objective* uncertainty that cannot be eliminated by eliminating interference. The technical way to describe this is that the statistical state of the decohered system is a mixture, but an *improper* one. The state of the marbles was a *proper* mixture. We need a proper mixture in order to say that we have solved the measurement problem, but decoherence does not provide that.

Yet perhaps a more serious challenge for the overarching goal of the decoherence program to explain the emergence of a classical (determinate, non-interfering) realm from the quantum realm is found in the recent work of Chris Fields (2011). Fields shows that in order to determine from the quantum formalism which pointer observable "emerges" via decoherence, one must first specify the boundary between the measured system and the environment; i.e., one must say which degrees of freedom belong to the system being measured and which belong to the environment. But in order to do this, one must use information available only from the macroscopic level, since it is only at that level that the distinction exists; only the experimenters know what they consider to be the system under study. So it cannot be claimed that the macroscopic level naturally "emerges" from purely quantum mechanical origins. The program is circular because it requires macroscopic phenomena as crucial inputs to obtain macroscopic phenomena as outputs.[18]

Therefore, the decoherence program does not actually solve the measurement problem, due to the persistence of improper mixtures which cannot be interpreted as mere subjective ignorance of existing ("determinate") facts or states of affairs. Nor does it succeed in the goal of demonstrating that the classical world of experience arises naturally from the quantum level.[19] In later chapters it will be shown that TI can readily account for the emergence of a macroscopic realm from the quantum realm.

1.3.2 Many worlds interpretations

Many worlds interpretations are variants of an imaginative proposal by Hugh Everett (1957), which he called the "relative state interpretation." The basic core of Everett's proposal was simply to deny that any kind of "collapse" ever occurs, and that the linear, unitary[20] evolution of quantum state vectors is the whole story. He suggested that any given observer's perceptions will be represented in one branch or other of the state vector, and that this is all that is necessary to account for our experiences. That is, the observer will become correlated with the system he is observing, and a particular outcome for the system can only be specified *relative to the corresponding state* for the observer (hence the title).

However, most researchers were not satisfied with this as a complete solution to the measurement problem. For one thing, it did not seem clear what was meant by an observer being somehow associated with many branches of the state vector. A

[18] Technically, Fields' argument is independent of the scale of the phenomena; it shows that classical information must be put in to get classical information (such as the relevant pointer observables) out. But in practice, this information comes from the macroscopic level – i.e., the experimenters' choices concerning what they want to study. See also Butterfield (2011, p. 17) for why the decoherence program does not solve the measurement problem.

[19] It should be noted that Deutsch (1999) and Zurek (2003) have presented "derivations" of the Born Rule. However, these derivations are observer-dependent, based on the specification of a non-intrinsic, classical division of objects into "system" and "observer" (or measuring device). Thus these approaches provide a subjective or purely epistemic probabilistic interpretation, based on defining ignorance on the part of some conscious observer. In contrast, TI derives the Born Rule in a physical way, with probability being a natural interpretation of what are pre-probabilistic physical weights. Thus objective probability arises out of a specific physical entity in TI – the incipient transaction. TI's physical, as opposed to epistemic, approach to probability is appropriate to the interpretation of quantum theory as being about objective, rather than subjective, probabilities. Another way to put it: Zurek and Deutsch's approaches are *epistemic* motivations in the same way that Gleason's is a "mathematical motivation" (as characterized by Schlosshauer and Fine, 2003). Insofar as they presuppose the presence of a classical "observer," they show consistency of quantum probabilities with what such an observer would observe, rather than deriving the probabilities in terms of a physical referent. The handicap hindering such accounts is that they must work with state vectors as the only physical referent. They do not have a physical referent for the projection operators (incipient transactions) which carry the real physical content of objective probabilities in quantum theory.

[20] "Linear" means that the quantum state only appears in the first power, and "unitary" means that no physically or mathematically ambiguous "collapse" has occurred. My reference to a "state vector" rather than a "wave function" is the most general mathematical form of the quantum state: an element of Hilbert space.

variant proposed by Bryce DeWitt "took the bull by the horns" and asserted that these branches described actual separate worlds – that is, that the apparent mathematical evolution of the state vector into branches corresponded to an actual physical splitting of the world. This version of Everett's approach became known as the full-blown "many worlds interpretation."[21] (Perhaps not surprisingly, the MWI has become the basis for many science fiction stories – a good example being the episode "Parallels" of *Star Trek: The Next Generation* (seventh season) in which the character Worf finds himself "transitioning" between different possible Everettian worlds with differing versions of events.) Proponents of MWI rely on decoherence in order to specify a basis for the splitting of worlds – that is, to explain why splitting seems to happen with respect to possible positions of objects rather than, say, their momenta, or any other mathematically possible observable.

Other Everettians, who adhere to a version called the "bare theory," prefer not to subscribe to an actual physical splitting of worlds, but instead attribute a quantum state to an observer and describe that observer's mental state as branching. Adherents of the bare theory argue that consistency with experience is achieved by noting that a second non-splitting observer (call him Bob) can always ask the first observer (Alice, who is observing a quantum system) whether she sees a determinate result, and Alice can answer yes without specifying what that result is.[22] Thus, an observer's state will either split along with a previous observer (if he inquires what the particular result was) and each of his branches will be correlated in a consistent way with the first observer's branches; or it will not split, and the second observer will still receive a consistent answer, if he only asks whether the first observer perceived a determinate result (but does not ask what the specific result is).

However, Bub (1997) and Bub *et al.* (1997) have argued that this approach ultimately fails to solve the measurement problem. Their critique is rather technical, but it boils down to two essential observations. (1) It turns out that there is an arbitrariness about whether the first observer will report "yes" or "no" concerning the determinateness of her perceptions, and that the choice of "yes" can be seen as analogous to choosing a "preferred observable" – that is, a particular observable that is assumed to always have a value. But that assumption contradicts the original intent of the interpretation – it is supposed to be a "bare" theory, after all, with no additional assumptions necessary besides the linear, unitary development of the quantum state. (2) It is not enough for Alice to simply report that she perceived a determinate result: we commonly take ourselves not only to perceive something definite, but also to perceive *what* that thing is. Bub *et al.* argue that inasmuch as the

[21] Bryce DeWitt (1970).

[22] Technically, this is described as Alice being in an eigenstate of "determinate measurement result," even if she is not in an eigenstate of one particular result or another.

"bare theory" exhibits feature (1), it is not really so "bare" after all and actually resembles what they term a "non-standard" approach to interpreting quantum theory: that is, an approach in which something is added to the "bare theory" such as the stipulation that one observable is to be "preferred" over others, either in having an always-determinate value or at least in being a "default" for determinacy. (Bohm's interpretation, to be discussed below, is an example of a non-standard approach of this type, in that position is the privileged observable.) And, regarding (2): as Bub *et al.* point out, other "non-standard" approaches can give an account of how Alice could report not only that she had some definite belief about the result she observed, but what that result was. So, in their analysis, the bare theory falls short, both of actually being "bare" and of actually solving the measurement problem.

As for the DeWitt full-blown MWI version of the Everett approach, a major challenge is to explain what the quantum mechanical weights, or probabilities, mean if each outcome is actually *certain* to occur in some branch (world) or another. Doesn't the fact that something comes with a probability attached to it mean that there is some uncertainty about the actual outcome? The basic position of MWI – that all outcomes will certainly occur – has led to rather tortuous and esoteric arguments about the meaning of probability and uncertainty.[23]

But the situation may yet be worse for Everettian interpretations. Recently, Kent (2010) has pointed out that the whole program of deriving the Born Rule[24] from a decision-theory approach based on the presumed strategies of rational inhabitants of a "multiverse" (a MWI term for the entire collection of universes) may be suspect. Any presumed strategy of a "rational" agent is no more than that – a probably sensible strategy among other possibly sensible strategies, and is therefore not unique. As Kent (2010) puts it:

> The problem is that abandoning any claim of uniqueness also removes the purported connection between theoretical reasoning and empirical data, and this is disastrous for the program of attempting to interpret Everettian quantum theory via decision theory. If Wallace's arguments are read as suggesting no more than that one can consistently adopt the Born rule if one pleases, it remains a mystery as to how and why we arrived at the Born rule empirically. (p. 10)[25]

More straightforwardly, the essential point, as Albert (2010)[26] has noted, is that there is a big difference between arguing that it can be considered rational to behave as though the world were a certain way and that the world actually *is* that way. Many

[23] As Peter Lewis (2007) notes, "Greaves (2004, pp. 426–7) suggests giving up the assumption that a subjective probability measure [the weights appearing in the set of possible outcomes] over future events requires uncertainty about what will happen, and Wallace (2006, pp. 672–3) suggests giving up the assumption that uncertainty requires some fact about which one is uncertain."

[24] The Born Rule is the prescription for calculating probabilities; see note 13.

[25] Kent refers to Wallace (2006).

[26] Albert, D. (forthcoming) (as referenced in Kent, 2010, p. 10).

of the approaches to justifying the Born Rule in Everettian theories depend on assumptions about what a rational agent would do, and on assumptions about mind–brain correspondences which are highly speculative as well as explicitly dualistic. As Kent (2010) observes:

> . . . the fact that we don't have a good theory of mind, even in classical physics, doesn't give us a free pass to conclude anything we please. That way lies scientific ruin: any physical theory is consistent with any observations if we can bridge any discrepancy by tacking on arbitrary assumptions about the link between mind states and physics. (p. 21)

Nevertheless, it would seem that Everettian arguments for the emergence of the Born Rule are crucially based on just such assumptions.

1.3.3 Bohm's interpretation

In a nutshell, David Bohm (1917–1992) proposed that the measurement problem can be solved by adding actual particles, possessing always-precise positions, to the wave function. To distinguish these postulated objects from the general term "particle" which is often used to refer to a generic quantum system, I will follow Brown and Wallace (2005) in terming these postulated Bohmian objects "corpuscles." The "equilibrium" distribution of these corpuscles is postulated to be given by the square of the wave function, in accordance with the Born Rule. The uncertainty and indeterminacy discussed earlier is still present in the Bohmian account due to the uncontrollable disturbance of any measuring device's interaction with these corpuscles; thus, we cannot know what their positions were prior to detecting a particular measurement result. That is, the knowledge we can have of corpuscle positions at any time before a given measurement is limited to the distribution given by the square of the wave function of the system of interest (for example, an electron in a hydrogen atom) (see Figure 1.5). The wave function then acts as a guiding or "pilot wave" for the corpuscle, as first suggested by Louis de Broglie (1923).[27] At the end of a measurement, the wave function will still have various "branches" (corresponding to different possible outcomes), but the corpuscle will only occupy one of them, and according to Bohm's interpretation, this determines which result will be experienced. Thus the idea is that the Bohmian corpuscle acts as a kind of "agent of precipitation" which allows for the experience of one outcome out of the many possible ones. In terms of measurement, Bohm argues that the "corpuscular" aspect of the measuring apparatus, on interacting with the measured quantum system, ultimately enters one of

[27] As far as I know, there is no physical account of how the "guiding wave," which lives in a $3N$-dimensional configuration space (where N is the number of corpuscles), guides the corpuscle – which is postulated to live in physical space. In the interest of a "level playing field" for competing interpretations, this lacuna should be kept in mind when considering criticisms of TI asserting that no specific "mechanism" is given for how a transaction forms or is actualized.

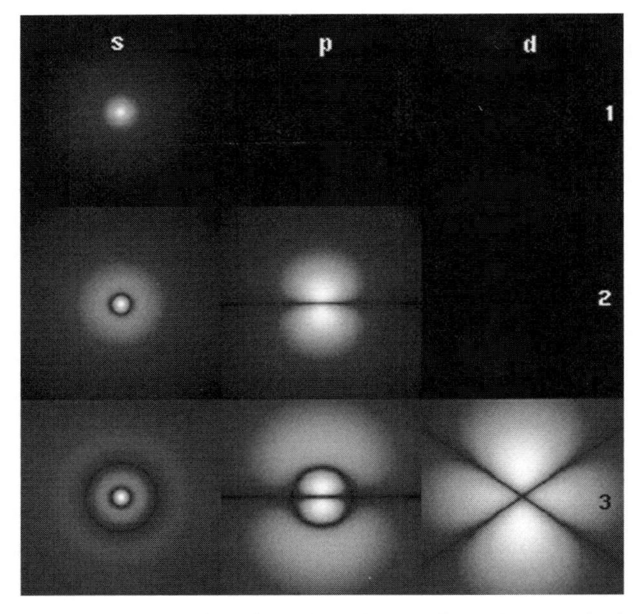

Figure 1.5 The squared wave function of an electron in various excited states of the hydrogen atom.
Source: http://en.wikipedia.org/wiki/File:HAtomOrbitals.svg

the distinct guiding wave "channels" of the wave function of the entire system (apparatus plus quantum system) created through the process of measurement, and this process singles out that particular channel as the one which yields the actual result. (Brown and Wallace call this the "result assumption."[28])

[28] Brown and Wallace, in their careful analysis of Bohm's seminal 1952 papers, comment in passing that Bohm apparently did not intend to "surpass" quantum theory – to propose, in their words, a theory with "truly novel predictions" (Brown and Wallace, 2005, p. 521). This may be a reference to the fact that the Bohmian approach amounts to a slightly different theory from standard quantum theory (cf. Valentini, 1992). The aspect of concern to me is the characterization of such a development as a "surpassing" of quantum theory and the implication that a good interpretation should make "novel predictions" (i.e., predictions that deviate from those of standard quantum theory). This language seems to imply that quantum theory is in need of improvement or remediation, and that a proper interpretational approach should generate a "better" (different) theory. In contrast, I think nothing is wrong with the theory itself and that prevailing interpretational approaches have not gotten to the root of the measurement problem: namely, the need to include absorption as a real physical process generating advanced states (confirmations). I do not believe that a successful *interpretation* needs to generate any novel predictions, but merely to provide a coherent and illuminating account of the theory itself, which effectively addresses the measurement problem. As a side note, an anonymous referee once commented in response to a statement like the preceding: "Since when has physics *not* dealt with difficult interpretational problems by changing the theory?" However, such changes were made not in response to *interpretational* problems, but rather to deal with the failure of a particular theory's *predictions*. For example, classical electrodynamics prior to relativity predicted that the speed of light should be dependent on the observer's motion. This prediction was refuted by the Michelson–Morley experiment. In contrast, the predictions of quantum theory are impeccable; it is probably the most strongly corroborated modern physical theory we have. What is at issue is arriving at a proper understanding of why the theory has the structure that it does. To modify the theory, I believe, is to fail to address the real scientific challenge it presents: what unexpected message does it convey about reality?

1.3.4 von Neumann's projection postulate

The formulation of John von Neumann, one of the pioneers of measurement theory in quantum mechanics, is not so much an interpretation as an analysis of the logical and statistical characteristics of the theory. It was von Neumann who first realized that the mathematical structure of the theory is a special kind of vector space (called a *Hilbert space*, in honor of David Hilbert who first defined it). While systems in classical mechanics can be represented mathematically as simple points labeled by their spatial position and momentum (technically, their coordinates in "phase space"), quantum systems have to be represented by rays in Hilbert space, which are objects that do not have simple coordinate-type labels, and which reflect an infinitely expansive ambiguity as to the "actual" characteristics of the systems they represent. Roughly speaking, one can think of the classical phase space coordinatization as only one of an infinite number of ways to provide a coordinatization in Hilbert space.[29]

Von Neumann's view of measurement is often referred to as "the standard collapse approach," since it simply assumes that, on measurement, the state of the quantum system "collapses" in a particular way (technically, it is "projected" onto a particular state corresponding to the type of measurement performed). He identified two different types of processes undergone by quantum systems: the "collapse" or "projection" that occurs on measurement he termed "Process 1"; and the simple deterministic evolution of a system's state between measurements he termed "Process 2." Of course, he left unclear exactly what is supposed to precipitate the collapse of "Process 1," and this remains part of the measurement problem. (An additional problem traditionally associated with collapse is that it appears to be in conflict with relativity, since it seems to call for a preferred frame of simultaneity denied by relativity. On the other hand, TI's approach to collapse is harmonious with relativity, as will be demonstrated in Chapter 6.)

As I will discuss later in the book, the question of what triggers collapse cannot be properly answered unless absorption is included in the dynamics. Without it, there is no clear "stopping point" at which a measurement can be regarded as completed (this was alluded to in Section 1.2.3), and all we have are vague "irreversibility" arguments that attribute *apparent* collapse to environmental dissipation or to "consciousness," but never really allow for a genuine physical collapse. At some point, an arbitrary "cut" is made at which the measurement is declared finished, "for all

[29] This observation reinforces the point made in note 28: the mathematical structure of the theory is qualitatively different from that of classical mechanics, in a very striking way. To understand the physical reason for this mathematical structure, I suggest, is the real interpretational challenge. The Everettian approach is one way of embracing the challenge, but I think it fails because it disregards half the dynamics (the advanced solutions to the complex conjugate Schrödinger equation) and cannot provide a physical (as opposed to epistemic/statistical) explanation for the Born Rule.

practical purposes" (a phrase which is often abbreviated "FAPP" in honor of John Bell who introduced the term as an expression of derision[30]). This arbitrary demarcation between the microscopic systems clearly described by quantum theory and the macroscopic objects which "measure" them is often referred to as the "Heisenberg cut" in view of Heisenberg's discussion of the issue (cf. Bacciagaluppi and Crull, 2009).

Under TI, with absorption taken into account, collapse occurs much earlier in the measurement process than is usually assumed, so that we don't need to include macroscopic objects such as Geiger counters, cats, or observers in quantum superpositions. This aspect is discussed in Chapters 3 and 4.

1.3.5 Bohr's complementarity

Neils Bohr, one of the pioneers of quantum theory along with Werner Heisenberg, developed a philosophical view of the theory which he termed "complementarity." I will not pretend to provide a detailed account of this view, which has been the subject of enormous quantities of research and elaboration, but I note that it is Kantian in character. (Kant's views will be described in detail in Chapter 2.) Bohr considered the properties of quantum systems to be fully dependent on what observers choose to measure, in that the experimental setup determines what sorts of properties a system can exhibit.[31] The Kantian flavor of his approach consists in denying that it is even meaningful to talk about the nature of the systems "in themselves," apart from their being observed in a macroscopic context. Based on Bohr's designation of such questions as "meaningless" or as beyond the domain of legitimate inquiry, his approach has been sardonically referred to as "shut up and calculate" (SUAC), a phrase coined by David Mermin (1989).

1.3.6 Ad hoc non-linear "collapse" approaches

So-called "spontaneous collapse" approaches such as that first proposed by Ghirardi, Rimini, and Weber (GRW) (Ghirardi *et al.*, 1986) impose an explicit theoretical modification on the mathematics of the standard theory – an additional non-linear term in the usual dynamics – in order to force a collapse into a determinate state. The added non-linear component takes a poorly localized wave function and compresses it. This approach is explicitly and unapologetically ad hoc and faces several problems, among them the following. (1) A wave function which is

[30] Bell introduced this term in his essay, "Against measurement" (Bell, 1990).

[31] Bub has shown (Bub, 1997) that complementarity can be viewed as a kind of "preferred observable," "no-collapse" approach, akin to the Bohmian interpretation which views position as the preferred observable. Bohr's preferred observable is whatever is measurable using the experimental setup.

compressed in terms of position must, by the uncertainty principle, gain a large uncertainty in momentum and therefore energy, which opens the door for observable effects, such as a system suddenly heating up – such effects are never observed. (2) Such collapses could only occur rarely, otherwise the well-corroborated normal evolution of the wave function would be noticeably disturbed. So it is not clear that their occurrence would be sufficient to account for the determinate results we see. Such "compression of the wave function" approaches are generally acknowledged as not viable, even by proponents of non-linear collapse, and Tumulka (2006) has proposed a variant which purports to avoid some of the pitfalls known to afflict the original GRW approach.

Tumulka's proposal, a "relativistic flash ontology" version (rGRWf), avoids the compression problem (1) cited above. However, rGRWf still involves a physically unexplained and ad hoc "collapse" mechanism, and evades what I believe is the central interpretational issue of explaining why the theory has the mathematical Hilbert space structure that it does (see notes 28 and 29). In addition, in order to be reconcilable with relativity, rGRWf ultimately appeals to time symmetry. TI already makes use of time symmetry without needing to make any ad hoc change to the basic theory. I deal with this issue in more detail in Chapter 6.

1.3.7 Relational block world approaches

The term "block world" refers to a particular kind of ontology[32] in which it is assumed that spacetime itself exists as a "block" consisting of past, present, and future events. The block is unchanging and it is only our perception of it that seems to involve change as we "move" along our worldline. Such a view seems implied by relativity, and some researchers have proposed that quantum theory should be interpreted against such a backdrop. The challenge in doing so lies in explaining why the unitary evolution of a particular quantum state "collapses" to a particular result. Adherents of this view propose that such events simply correspond to a discontinuity of the relevant worldlines: that it is just a "brute fact" about nature that such discontinuities must exist.

This principle of a spacetime block with uncaused (primal) discontinuities was pioneered by Bohr, Mottelson, and Ulfbeck (BMU), who say (Bohr *et al.*, 2003):

> . . . The principle, referred to as genuine fortuitousness, implies that the basic event, a click in a counter, comes without any cause and thus as a discontinuity in spacetime. From this principle, the formalism of quantum mechanics emerges with a radically new content, no longer dealing with things (atoms, particles, or fields) to be measured. Instead, quantum

[32] "Ontology" refers to what is assumed to exist; what is real.

mechanics is recognized as the theory of distributions of uncaused clicks that form patterns laid down by spacetime symmetry. (abstract)

BMU take macroscopic "detector clicks" as primary uncaused events and refer to atoms as "phantasms." Thus they are explicitly antirealist about quantum objects. BMU's approach has been developed more recently into a "relational block world" (RBW) interpretation by Silberstein, Stuckey, and Cifone (Silberstein *et al.*, 2008). RBW advocates take spacetime relations and their governing symmetries as fundamental and attempt to derive a version of quantum mechanics based on this ontology.[33] One basis for criticism of RBW is that it makes fundamental use of dynamical concepts such as momentum while denying that those concepts refer to anything dynamical.[34]

1.3.8 Statistical/epistemic approaches

Some researchers (e.g., Spekkens, 2007) have been investigating an approach in which the quantum state reflects a particular preparation procedure but does not necessarily describe the physical nature of the quantum system under study. This implies that the quantum state characterizes only our knowledge; "epistemic," from the Greek word for "knowledge," is the technical term used. The statistical aspect consists in connecting a particular preparation procedure to a particular distribution of outcomes. The key feature distinguishing this "statistical" approach from the "hidden variables" approaches – such as Bohm's theory – is that in the former the quantum state is not uniquely determined by whatever "hidden" properties the quantum system possesses. In contrast, a quantum system under the Bohm theory is physically described by its wave function as well as an unknown position x of the postulated particle associated with the wave function; there is only one wave function that can be associated with these properties, even though the same wave function can be associated with another system with a different particle position x'.

A new theorem by Pusey *et al.* (2011) casts serious doubt on epistemic/statistical approaches. It shows that, given some fairly weak assumptions, the statistics of a system whose state is not uniquely determined by its physical properties can violate the quantum mechanical statistical predictions.[35] The implication is that the

[33] I do, however, share RBW's rejection of a "building block" ontology: the empirical world is a network of transactions, not collections of primitive individuals.

[34] For example, in RBW, experimental configurations are described by symmetry operators such as the translation operator $T(a) = \begin{pmatrix} e^{-ika} & 0 \\ 0 & e^{ika} \end{pmatrix}$, because momentum k is the generator of spatial translations. But, in RBW, there are no entities that possess momentum. It thus remains unclear what dynamical terms such as "momentum" refer to, in an adynamical account such as RBW.

[35] Granted, one of those assumptions is that there is no retrocausality. However, it is unclear to what extent adding retrocausality about an underlying ontology would help to support the basic statistical/epistemic program, which

quantum state really does describe a physical system, not just our knowledge of our preparation procedure.

1.4 Quantum theory presents a genuinely new interpretational challenge

Some researchers take the point of view that the appropriate response to quantum theory's apparently intractable puzzles is to adopt a strictly empiricist, pragmatic point of view, for example to simply say that there is no physical explanation for the puzzling behavior of quantum objects as reflected in the theory, that nature simply "refuses to answer" the questions we try to pose about that behavior. This assumption could be seen as a version of the Bohrian/Kantian view that people can gain knowledge *only* of the phenomenal level of appearance; that quantum theory might permit us to "knock at the door" of the sub-empirical, sub-phenomenal world but that the door must remain forever closed. This approach, I believe, is to evade a genuine, non-trivial interpretational challenge posed by the theory; i.e., it admonishes us to renounce the realist approach of assuming that physical theories can describe nature itself, at all levels.

While I certainly agree with the idea that quantum theory has an unexpected message, I think that message *is* one about reality – like all profoundly corroborated and powerfully predictive theories – and that the challenge is to figure out what the theory is telling us about reality. As this book will reveal, I think it is an exciting, strange, and indeed revolutionary message; certainly more interesting and revolutionary than the notion that theories of small things can only be about subjective knowledge or only about appearances.

It was the behavior of hydrogen atoms that inspired Heisenberg to arrive at his first successful version of quantum theory. Clearly the theory he arrived at was about those atoms and not just about his knowledge, since without reference to, and guidance from, those atoms he would never have constructed the theory. That is, the theory's structure was *driven by the behavior of atoms*. Yes, the "observable behavior" of atoms, but the conclusion that the theory is only about our knowledge of them does not follow (and this point will be explored further in the following chapter).

The true puzzle of quantum theory is that there are physical entities beyond our power to perceive directly in the ordinary way, and that they behave in strange and amazing ways. This is not just anthropocentrically about "our knowledge," it is also about them. What are they saying to us? Heisenberg listened, and in the next chapter I will further explore his initial insights.

is to restore a more commonsense (i.e., classical) interpretation of quantum states than appears to be available from being realist about quantum states. If one is going to admit retrocausal influences anyway, then why not embrace a straightforward realist time-symmetric interpretation such as TI?

2

The map vs. the territory

> [Quantum] theory is so rich and counterintuitive that it would not have been possible for us, mere mortals, to have dreamt it without the constant guidance provided by experiments. This is a constant reminder to us that nature is much richer than our imagination.
>
> *Jeeva Anandan (1997)*

In this chapter, I consider some general issues of interpretive methodology, to present to the reader the motivation behind the new TI. I first acknowledge the proposed interpretation as applying to a functioning, non-idealized theory; that is, neither the original TI, nor the current proposal, is a "rational reconstruction" of quantum theory. I then argue in favor of a realist approach as opposed to an instrumental one.[1]

2.1 Interpreting a "functioning theory"

The present work offers an interpretation of what MacKinnon (2005) calls a "functioning," or informal theory: non-relativistic quantum mechanics and its extension into the relativistic domain via quantum field theory.[2] Since functioning theories are often inherently "untidy" (either in a mathematical or conceptual sense or both), philosophers of physics often engage in "rational reconstruction" of theories in order to render them more logically self-consistent in the hopes that the resulting formal theory will better lend itself to an unambiguous interpretation. However, as MacKinnon (2005) observes, history does not support the notion that

[1] This chapter primarily addresses instrumentalist views; however, many so-called "realist" approaches to quantum theory contain unacknowledged instrumentalist or positivist-flavored assumptions about what the term "reality" means (such as "real" = "empirically detectable"), so the discussion herein is relevant to those as well.

[2] As an example of this "untidiness," non-relativistic QM and its relativistic extension might well be considered two different functioning theories, yet clearly they must describe the same reality and therefore presumably must be parts of a larger theory. A point of contact is found in Zee's observation (Zee, 2010, p. 19) that non-relativistic quantum mechanics can be obtained in the Lagrangian formulation as a 0+1-dimensional quantum field theory.

such recast, formalized theories lead to robust ontological insights.[3] He instead characterizes the interpretive task as one of "find[ing] a way of relating philosophical questions about epistemology and ontology to functioning physical theories, rather than idealized constructions" (p. 4). That, in a nutshell, is the aim of the present work.

2.2 The irony of quantum theory

The original inception of quantum theory and the course of its subsequent evolution contain a deep irony. To appreciate this irony, we first need to revisit a bit of history.

2.2.1 Heisenberg's breakthrough

A major breakthrough in quantum theory was achieved in 1925 through a decision by German physicist Werner Heisenberg to let go of certain preconceived metaphysical assumptions about the nature and behavior of matter: specifically, that we could picture electrons as little particles – corpuscles in the Greek (Democritan) conception – orbiting an atomic nucleus. Facing a theoretical impasse in accounting for atomic phenomena, he renounced these classical *anschaulich* (German for "picturable")[4] assumptions and retained only observable quantities such as energy differences and radiation frequencies, which could be measured and recorded as hard data. These he entered into arrays which he sardonically termed "laundry lists," and which his then-teacher Max Born would soon realize were matrices (arrays of numbers in a form well known in mathematics). Thus was born Heisenberg's "matrix mechanics" version of the theory, which successfully predicted the experimental (spectral) data arising from observations of the hydrogen atom. Subsequent development would eventually lead to a powerful, empirically successful theory which could be expressed in different forms (probably the best known being the Schrödinger wave mechanics, based on Erwin "The Cat" Schrödinger's celebrated equation), and whose formal structure was described, as von Neumann had first noticed, by an abstract mathematical space called Hilbert space.

[3] He cites, as an example, Maxwell's brilliant unification of electricity and magnetism by way of the "electric displacement current," which was subsequently not regarded as having fundamental ontological content but rather as making possible the formalization of the theory as a unified set of equations (Maxwell's equations).

[4] The term *anschaulich* presupposes that "picturable" means the usual classical picture of corpuscles following determinate trajectories. This assumption is contested in the present account: physical processes could be "picturable" in terms of an entirely different kind of picture.

2.2.2 Bohr's antirealism

However, as observed in Chapter 1, nearly a century later researchers are still deeply puzzled about how to interpret the theory, in the sense of understanding what it says about reality (if anything). Most physicists and philosophers of physics are aware that Heisenberg's breakthrough came as a result of renouncing his preconceived metaphysical assumptions; and many of them (including, most notably, Heisenberg's fellow quantum theory founder Niels Bohr) have taken from this fact what I believe is the wrong lesson: they have renounced realism with regard to quantum theory. That is, the idea that there was some understandable, underlying physical reality described by quantum theory tended to be viewed suspiciously, as a misguided impulse to drag in metaphysical baggage that Heisenberg's approach had discredited as inappropriate methodology. Probably nobody says this more emphatically than Neils Bohr: "There is no quantum world. There is only an abstract physical description. It is wrong to think that the task of physics is to find out how nature is. Physics concerns what we can say about nature."[5]

The last sentence by Bohr assumes that we can only talk about nature in terms of classical concepts, i.e., the very "picturable" notions that Heisenberg had renounced in order to arrive at his matrix formulation of quantum theory. Bohr viewed such concepts as indispensable for communicating experimental results and, in general, for talking about physical reality. In fact, he elevated this claim to the level of a fundamental epistemological principle. Bohr's positivistic prohibition on "finding out how nature is" was not necessarily heeded by everyone, but it had, at the very least, a chilling effect on interpretive inquiry.

Bohr's legacy is alive and well among many practicing physicists, whose job it is to calculate experimental predictions and analyze results, and who tend to regard efforts by philosophers of physics to "find out how nature is" to be a misguided waste of time. Many of them approach interpretational puzzles of quantum theory from the kind of deflationary, "debunking" view alluded to at the end of the previous chapter. Of course, nobody is to be faulted for choosing not to be realist about physical theory, especially when it is not in their job description to do so. But the main thesis of this work is that, contra Bohr, it is perfectly *reasonable* to be realist about the subject matter of quantum theory, and that it is perfectly possible to "find out how nature is," as long as we don't expect it to be "classically *anschaulich*" and are willing to entertain some new and apparently very strange ideas of how nature might be (analogous to the strange specter of energy having to be "quantized" which

[5] As quoted in *The Philosophy of Niels Bohr* by Aage Petersen (1963).

accompanied Max Planck's successful derivation of the blackbody radiation spectrum).[6]

2.2.3 Einstein's realism and a further irony

Einstein, as is well known, completely disagreed with Bohr's approach. His motivation was, in his own words, to "know God's thoughts."[7] Yet, ironically, a similar antirealist tendency has recently arisen based on the methodology Einstein used in formulating his theory of special relativity. Einstein famously arrived at his theory by thinking in terms of what someone could actually measure with (idealized) rigid rods and clocks, and concluded that one needed to renounce certain metaphysical notions about space and time: in particular, Newton's view that space and time are absolute, immutable "containers" for events. What is less often remembered is that Einstein also used formal theoretical assumptions: in particular, he demanded the invariance of electromagnetism, requiring that the theory not be dependent on an observer's state of motion. But the prevailing message of relativity came to be that there is no such thing as absolute simultaneity or absolute lengths of objects, and that these concepts were metaphysical ballast to be jettisoned. Einstein's renunciation of such absolute metaphysical concepts is often amplified, like Heisenberg's renunciation of the trajectory concept, into a universal doctrine that any notion of an underlying (i.e., sub-empirical) reality is to be eschewed.

However, not only is this an inappropriate lesson to take from these theoretical achievements, it is not even consistently applied: most researchers (and especially physicists) continue to be thoroughgoing realists about spacetime, viewing it as a fundamental substantive "container" or backdrop which not only underlies all possible theoretical models but which even has causal powers to "steer" particles on trajectories.[8] (And note the additional irony that the notion of "trajectory" is still very much with us despite the prevailing view that fundamental reality should not be considered "picturable."[9])

[6] Planck had introduced a discrete sum of finite energy chunks as a calculational device only. When he tried to take the limit of the sum as the size of the chunks approached zero, he got back the old – wrong – expression. The chunks had to be of finite size in order to get the correct prediction.

[7] "I want to know God's thoughts. The rest is details." Widely attributed to Einstein.

[8] The commonplace notion that spacetime has causal power to steer particles is subject to sustained and cogent criticism by Harvey Brown (2002).

[9] For example, many discussions of the "two-slit" experiment and similar experiments, in which the state of a single quantum is placed into a superposition by a half-silvered mirror or other means, are centered around so-called "which-way information." This term is heavily laden with the presumption of a determinate trajectory: surely, if one talks about "which-way information," one tacitly assumes that the entity under study went either one (spacetime) way or the other; i.e., pursued a trajectory. So, even though perhaps not always intended, its use smuggles in a supposedly renounced classical metaphysical picture.

2.2.4 *Theory construction vs. theory interpretation*

The point generally overlooked in the trend described above is that theory formulation/discovery is an entirely different process from that of theory interpretation. We need to distinguish between (i) the valid point that preconceived metaphysical assumptions can serve as a barrier to theory *invention or discovery*, especially when a successful new theory cannot be based on such assumptions; and (ii) realist *interpretation* of an existing empirically successful theory as a way of discovering *new* features of reality uncovered by that theory. The deep irony of quantum theory, I suggest, is that its discovery was made possible by the renunciation of a then-realist approach and attendant metaphysical baggage; yet when interpretationally queried from a realist perspective in the proper way, quantum theory can open the way to an entirely new and richer understanding of physical reality: a strange new kind of model that we could not have discovered without first letting go of inappropriately classical metaphysical concepts. In making this claim, I invite the reader to reflect on the insightful quote by the late Jeeva Anandan which began this chapter.

2.3 "Constructive" vs. "principle" theories

What do I mean by querying a theory "in the proper way"? In order to address this, I first need to review an important distinction in theory type: "constructive" vs. "principle" theories. Simply put, a constructive theory is one based on a model. A famous example is the kinetic theory of gases, which represents the behavior of gases in terms of small, impenetrable spheres in collision with one another and the walls of their container. By applying known physical laws to this model, James Clerk Maxwell and Ludwig Boltzmann were able to deduce the large-scale thermodynamic behavior of gases; for example, Boyle's Law relating temperature, pressure, and volume ($PV = nRT$).[10] Such a "constructive" theory is powerful and illuminating because it allows us to understand the "nuts and bolts" of what is really going on at a level beyond ordinary experience, i.e., beneath the phenomenal level of appearance. That is what Einstein meant when he talked about wanting to "know God's thoughts." He didn't just want to know about how God's creation *appears* and to be able to analyze, classify, and predict those appearances; he wanted to know how it all works beneath the merely phenomenal level, "to boldly go" where Bohr summarily pronounced that nobody should be able, nor wish, to go.[11]

In contrast, a "principle" approach to theory development lacks a physical model. It starts from an abstract principle or principles that serve to constrain the form that the theory can take, and then fits the theory, with the help of mathematical

[10] A comprehensive and very readable account of this scientific episode is found in Brush (1976).
[11] With apologies to Gene Roddenberry.

consistency and basic physical laws such as energy conservation, to empirical observation. Relativity was a principle theory, and Einstein was very dissatisfied with this aspect of it. He felt that only a constructive theory, with its attendant illuminating model, provided genuine insight into "how nature really is." Similarly, quantum mechanics was a principle theory, as we can see by the fact that Heisenberg had to explicitly jettison the models he was trying to work with (i.e., his erroneous metaphysical pictures of how electrons behaved), and to work only with empirical observations which served to constrain the form of the theory. Before that, Planck used a purely mathematical trick – summing over discrete energy levels instead of assuming energy was a continuously variable quantity – to obtain the correct empirical result for blackbody radiation (see, e.g., Eisberg and Resnick, 1974, section 1.1 and especially p. 14 for a clear account of how this phenomenon presented a fatal problem for classical electromagnetism and forced the invention of quantum theory). His desperate resort to this tactic led to the discovery of Planck's constant h, the fundamental physical constant which characterizes the smallest unit of action (units of energy times time or momentum times length). Thus his approach to the discovery of the new theory was principle-based (i.e., using formal mathematical considerations), not model-based.

2.4 Bohr's Kantian orthodoxy

Now, as noted above, Bohr was perfectly content with the idea that quantum mechanics was a principle theory. He assumed from the way that the theory was arrived at – by rejecting a model that didn't work – that there *can be no model* for quantum theory, i.e., no way of picturing "how nature is." In other words, Bohr elevated the fact that one cannot apply *classical* model-making to a *non-classical* realm into a broad-brush policy that, at the quantum level, one should not try to find models of *any* kind. He basically claimed that if one cannot have a classical model, there can be no model, and that quantum theory represents the end of the scientific search for understanding of the physical world in a realist sense: i.e., independently of how we happen to be looking at it.

Put differently, he assumed that classical modeling is *equivalent* to giving a realist account of micro-reality, and that one therefore cannot give such an account. He repeatedly pointed out that, in order for scientists to communicate their results and thereby establish objective (or at least intersubjective) accounts of phenomena, they had to be able to talk about determinate pointer readings (and thereby speak in "classical language"). This is true, of course – it is the means by which all physical theories are tested and corroborated. Nevertheless, the physical content of quantum theory is not necessarily exhausted by its empirical correspondence. The interpretational question is whether the additional formal content (e.g., the Schrödinger equation and its solutions) has some physical referent, regardless of whether or not that physical referent can be directly observed (or even understood in macroscopic terms).[12]

At this point it is useful to acknowledge a distinct similarity between Bohr's thought and the work of the great German philosopher Immanuel Kant. Kant proposed that reality has two fundamental aspects: (1) the world of appearance and (2) the "thing-in-itself" (or "noumenon"), which he held was unknowable.[13] (For an accessible introduction to the problem of gaining knowledge of the "thing-in-itself," the reader is encouraged to consult chapter 1 of Bertrand Russell's *The Problems of Philosophy*, in which the author considers an ordinary table and presents a convincing case that the table itself, apart from any perception of it, is a deeply mysterious object, if it even exists at all. For an updated version of this epistemological puzzle, see Section 7.5.) Kant also proposed that there are "categories of experience" that make knowledge of the world of appearance possible, and which are the only means through which knowledge is constructed.[14] (Knowledge, for Kant, was *only* about (1) the world of appearance; recall that part of the definition of (2), the thing-in-itself or underlying reality, was that it was intrinsically unknowable.) Among the "categories of experience" were concepts like space, time, and causality. In particular, Kant proclaimed that Euclidean space was an *a priori* category of understanding, meaning a necessary concept behind any knowable phenomenon – an assertion which, it should be noted, has since been decisively falsified by relativity's non-Euclidean accounts of spacetime.

Bohr seems to have assumed, much like Kant, that all knowledge obtained by way of physical theories applies only to the world of appearance and that the "classical modes of description" are required for all knowledge. So Bohr's "classical modes of description" play the same role as Kant's "categories of experience." Bohr, in essence, proclaimed that while quantum theory might have placed us just at the doorstep of the "noumenal" realm, the nature of the theory required that we could not gain knowledge about it and that, moreover, it would be scientifically and methodologically unsound to think that we should try to do so, as reflected in his previous quote. By "abstract quantum mechanical description," Bohr pre-emptively denied that the formalism could be referring to anything physically real, and considered it only a linguistic or computational device. I believe that this assumption can and should be questioned.

[12] Ernan McMullin, as quoted in Ladyman (2009), makes this point quite clearly: "[I]maginability must not be made the test for ontology. The realist claim is that the scientist is discovering the structures of the world; it is not required in addition that these structures be imaginable in the categories of the macroworld."

[13] Kant often used the "thing-in-itself" interchangeably with the term "noumenon," a Greek term which translates roughly as "object of the mind." Kant's division is very similar in structure to Plato's division, as the reader will recall from Chapter 1.

[14] Kant's ideas discussed here were presented in his *Critique of Pure Reason* (1996).

It has often been pointed out (e.g., by Bohr and Heisenberg) that in general there can be no mechanistic, deterministic account of individual microscopic events. This fact is often referred to in terms of "quantum jumps" that cannot be predicted, even in principle. Yet a realist understanding of micro-reality *need not* take the form of a detailed mechanical account of an individual event – the entity that remains elusive to causal description as Anton Zeilinger notes in his philosophical analysis.[15] To assume, like Bohr, that a realist understanding must be in terms of the usual "classical," causal account is to limit ourselves to a pseudo-Kantian "category of experience" which is shown to be obsolete by scientific advance, much as Kant's own prescribed "categories" became obsolete when (for example) it was discovered that theories of spacetime had to allow for non-Euclidean forms. The new realist understanding may not be in terms of causal, mechanistic processes. It may instead encompass a fundamental indeterminism at the heart of nature, but one which is well-defined in terms of the conditions under which it occurs – in contrast to prevailing "orthodox" interpretations which suffer from an ill-defined micro/macro "cut" (as discussed in Section 1.3.4). The new understanding offered here is a rational account, in the sense of being well-defined and self-consistent, even while it lacks certain features, such as determinism and mechanism, that have been traditionally assumed to be requirements for an acceptable scientific account of phenomena.

Thus, as alluded to above, I regard Bohr's conclusion as a logical fallacy: specifically, an overgeneralization. It simply does not follow logically that the failure of a particular kind of model entails that no model of any kind is possible.[16] (Alternatively, as above, one may regard Bohr as making the same kind of mistake as Kant when the latter presumed that there can be no knowledge of a realm that is not based on a Euclidean space.) While it may be true, as a matter of contingent fact, that there is no adequate model, I see no reason that a failure of a particular sort of inappropriate model should be turned into a general prohibition against modeling. On the contrary, I suggest that a principle theory can provide truly groundbreaking insights into new aspects of reality: that it can ultimately lead us to a *new* kind of model, one so utterly different from how we are used to thinking about reality that we could not have approached it directly, "from the ground up" so to speak, but had to arrive at it through an indirect route, "top down," as Heisenberg did.

[15] http://www.quantum.at/fileadmin/zeilinger/philosoph.pdf

[16] I recognize that Bohr adduced Kantian epistemological reasons for his prohibition against modeling in quantum theory, but I reject those as well. Specifically, it will be argued later on in this chapter that the promise of quantum theory is to give us a glimpse of the "noumenal" realm, so I will be rejecting the Kantian claim that all knowledge must be restricted to phenomena.

2.5 The proper way to interpret a "principle" theory

So, what is the "proper way" to interpret such a principle theory, one that was developed without reference to any model? To answer this question, let's turn to a famous dictum by Bryce DeWitt, who presented it as the essential motivation for his development of the Everett interpretation into what became known as the many worlds interpretation:[17] "The mathematical formalism of the quantum theory is capable of yielding its own interpretation" (DeWitt, 1970).

I take this to mean that the formalism resulting from whatever methodology was needed to develop an empirically successful theory – especially a "principle" theory like quantum mechanics, which was not based on prior construction of a model – has features that may well point to heretofore hidden or unnoticed features of reality. (A perfect case in point, again, is Planck's stumbling upon quantized energy because his theory said so, not because he wanted it that way.) Since the features of an empirically successful principle theory are (apparently) not something we could have thought of unaided, they are not available to us as a possible model, and we (like Heisenberg) have to proceed without their help, "groping in the dark," so to speak, aided only by previously established physical principles, mathematical consistency, and empirical data to guide us to the form of the theory.

Heisenberg, in choosing to "listen to reality" by renouncing his previous unhelpful metaphysical assumptions, wrote down the "laundry list" formalism (matrix mechanics) that turned out to be a useful instrument for predicting observations arising from the microscopic systems he was studying. But, as argued above, it does not logically follow that all there is to reality is those abstract "laundry lists." A possibly useful analogy here is a map to some buried treasure: Heisenberg, through his choice to adopt a Zen-like "beginner's mind" approach to the phenomena under study, stopped listening to his own ineffective ideas and began to listen to the message of reality instead, as encoded in the phenomena. Thus, he was able to "hear" what reality was trying to tell him by writing down what became a useful "map." The realist impulse that underlies and motivates all fundamental scientific advance is to acknowledge that there is some reason, however obscure, that such a theoretical "map" allows us to predict phenomena. Unless we wish to believe in miracles or coincidence as the explanation for the success of a theory like quantum mechanics (or to deny that theory success even needs explaining, which is to retreat from the deepest aspects of the scientific and philosophical mission), we are obligated to acknowledge that the "map" reflects something about reality – however utterly new and unfamiliar.

[17] As mentioned in Chapter 1, it is my view that MWI advocates overlook part of the formalism (advanced solutions).

Another analogy for the inspiration leading to a successful principle theory is in the realm of psychology and interpersonal relationships. Successful mediators know that conflicts can be resolved when the parties are helped to let go of their own preconceived notions, desires, or requirements for the other person, and start to listen to what he or she is saying. More broadly, a socially effective person has the ability to be receptive to the messages from his/her environment and the flexibility to adapt to the meaning of the messages, i.e., to let go of preconceived notions about "how things should be" and to behave in ways that are more appropriate and fruitful. But they don't conclude from that that there is nothing further to be learned about that other person or situation, or that there is nothing beyond those messages they heard which allowed them to behave more effectively. A new way of behaving is more "fruitful" because, to push the analogy of that adjective, there is *something there yielding fruit*. Heisenberg's approach exemplifies, albeit in a different context, the behavior of a successful person in social relationships. He stopped presuming and started listening, and was able to write down a very useful "map." We should not mistakenly conclude from his methodological success that there is no more to reality than that map.

So, as will be developed in later chapters, the "proper way" to interpret the theory is to "listen" carefully to its mathematical features. A crucial step was made by Max Born who linked the absolute square of the Schrödinger wave function[18] to something empirical, if only statistical: this quantity could be seen to function as the probability of observing that property when one conducted a measurement of the system. His finding became known as the "Born Rule," and it is the fundamental empirical link between quantum theory and the world of phenomena. As noted in the previous chapter, in most prevailing interpretations, the Born Rule is either simply assumed as part of the mathematical machinery that does not merit or require explicit interpretation, or it is given a pragmatic, "for all practical purposes" account which, in my view, fails to do it justice as the crucial link between theory and concrete experience. The Born Rule constitutes a deep mystery for all prevailing interpretations; there would appear to be no straightforward ontological (i.e., non-epistemic, non-statistical) explanation for it in any interpretation other than TI.[19]

[18] More generally, the probability is the square of the projection of the quantum state onto a particular classically observable property, e.g., position or momentum.

[19] As noted in Chapter 1, Bohmians claim that the Born Rule is obtained as the statistical distribution of particle positions. But this is only for the so-called "equilibrium state" of the subquantum level (i.e., the level of determinate positions). Since the Bohmian theory allows for the particle position distribution to deviate from the Born Rule, it is a slightly different theory from quantum mechanics. Even if one viewed the "non-equilibrium" state as improbable or even impossible, the account is only statistical, which I view as a weaker kind of physical explanation. A further challenge for the Bohmian account is that particles are continually created and destroyed in the relativistic regime, which would seem to increase the likelihood of distributions that might deviate from

2.6 Heisenberg's hint: a new metaphysical category

Heisenberg took a further step in "listening" to quantum theory when he made the following statement: "Atoms and the elementary particles themselves are not real; they form a world of potentialities or possibilities rather than things of the facts."[20] This assertion was based on the fact that quantum systems such as atoms are generally described by quantum states with a list of possible outcomes, and yet only one of those can be realized upon measurement. I think that he was on to something here, except that I would adjust his characterization of quantum systems as follows: they are real, but not *actual*. In his terms, they are something not quite actual; they are "potentialities" or "possibilities." Thus my proposal is that quantum mechanics instructs us that we need a new metaphysical category: something more real than the merely abstract (or mental), but less concrete than, in Heisenberg's terms, "facts" or observable phenomena. The list of possible outcomes in the theory is just that: a list of possible ways that things could be, where only one actually becomes a "fact." This proposal is directly analogous to Planck's proposal, in view of the inescapable formal features of his theory, that energy is quantized.

The distinction between a quantum possibility and a fact is clarified in a comment that Heisenberg made later in his life (and will be further clarified in Chapter 7):

> The probability wave of Bohr, Kramers, Slater ... was a quantitative version of the old concept of "potentia" in Aristotelian philosophy. It introduced something standing in the middle between the idea of an event and the actual event, a strange kind of physical reality just in the middle between possibility and reality.
>
> *(Heisenberg, 2007, p. 15)*

So, Heisenberg had arrived at a new kind of metaphysical understanding, a "picture," if you will, of the reality described by quantum theory. However, in view of his ambivalence about it – he was a practicing physicist, after all, and expected models to be based on "things of the facts" – he did not pursue this insight as a viable description of the underlying reality described by quantum theory. My goal in this work is to essentially pick up where he left off.

A further important aspect of "listening to the formalism" of quantum theory is to acknowledge its time-symmetric (or at least "advanced") aspects. Specifically, it cannot be overemphasized – since the fact is habitually neglected – that *advanced (negative-energy/time-reversed) states necessarily enter into any calculation needed to obtain empirical content* (i.e., probabilities for outcomes of measurements, or expectation values for the values of measured observables). Indeed, this

the "equilibrium" configuration needed for its empirical equivalence to standard QM. Many world or Everettian accounts give an epistemologically based account of the Born Rule which must refer to the knowledge of an observer.

[20] Heisenberg (1958, p. 186).

overlooked fact is so important that I will elevate it to an interpretational maxim for any realist interpretation:

Maxim: Mathematical operations of a theory which are necessary to obtain correspondence of the theory with observation merit a specific (exact) ontological interpretation.

This proposed maxim no doubt requires some elucidation. For one thing, TI's rival "purist" interpretation (that is, the collection of approaches constituting the so-called many worlds interpretation based on Hugh Everett's proposal of 1957) does not adhere to it. As alluded to earlier, MWI addresses the Born Rule by epistemological or statistically approximate methods: by arguing, via decision theory, that a rational observer would choose to bet on outcomes obeying the Born Rule; by arguing that Everettian worlds violating the Born Rule have approximately zero measure; etc. Similarly, the Bohm theory proposes that the distributions of Bohmian particles closely approximate that specified by the Born Rule. Now, in the absence of any mathematical property of the basic theory which could provide an unambiguous ontological basis for the Born Rule, such approximate and/or ad hoc approaches might be justified. But the theory *does* possess a specific mathematical object that can provide an exact ontological basis for the Born Rule: the set of advanced solutions which, under TI, are confirmation waves arising from the ubiquitous absorption processes neglected in other interpretations. Since absorption processes are physically present whenever there is a detection (the latter being a requirement for an observation), the advanced solution is the obvious mathematical entity to interpret as a component of the ontological basis for the Born Rule.

Furthermore, it should be noted that historically, Schrödinger began with a relativistic wave equation (now called the Klein–Gordon equation), but abandoned it when he saw that it had negative-energy (advanced) solutions.[21] As has been noted in Cramer (1986), the non-relativistic Schrödinger equation which ultimately proved successful is obtainable as a limiting procedure of that relativistic equation. The same limiting procedure leads *equally well* to the complex conjugate Schrödinger equation, which has negative-energy solutions. In standard interpretations, these solutions are ignored. TI simply proposes that they must be included for a solution of the measurement problem and a proper interpretation of the theory.

Despite the counterintuitive aspects of advanced states, I believe that truly hearing what the formalism is saying means taking seriously the idea that it describes something with advanced (as opposed to the usual retarded) qualities.

[21] Another issue with the Klein–Gordon equation is that it does not yield a positive definite probability density when given a single-particle interpretation. This is understandable since it is a relativistic equation, and particles are continually created and destroyed in the relativistic domain – so in any case it is inappropriate to view the Klein–Gordon equation as describing a stable single-particle state.

This is where, in my view, TI improves upon Everettian interpretations which try to approach the formalism from a receptive, "purist" point of view, but which fail to notice that the advanced states are a crucial part of the theory with physical content that should not be neglected.

The transactional conceptual picture represents a parallel to that of Einstein's conceptual unification of the instrumental and pragmatic pre-relativistic quasi-theories, as described by Zeilinger (2009):

> It so happened that almost all relativistic equations which appear in Einstein's publication of 1905 were known already before . . ., mainly through Lorentz, Fitzgerald and Poincaré – simply as an attempt to interpret experimental data quantitatively. But only Einstein created the conceptual foundations, from which, together with the constancy of the velocity of light, the equations of the theory of relativity arise. He did this by introducing the principle of relativity, which asserts that the laws of physics must be the same in all inertial systems. I maintain that it is this very fact of the existence of such a fundamental principle on which the theory is built which is the reason for the observation that we do not see a multitude of interpretations of the theory of relativity. (p. 2)

The Born Rule equating the probability of a particular result to the square of the wave function is one of the equations allowing quantitative interpretation of experimental data in quantum theory, just as the Lorentz contraction allowed quantitative empirical correspondence in pre-relativistic theories. The current multitude of competing "mainstream" interpretations of quantum theory (among these the Bohmian theory, "spontaneous collapse," or GRW approaches, MWI) are all different ways of providing approximate, pragmatic, after-the-fact justifications for the Born Rule and the conditions of its application – showing that its use is consistent with the rest of the theory in some limit – rather than an *explanation for how it arises* naturally from the theory. In contrast, the conceptual picture of a transactional process is what allows the operational equation of the Born Rule to arise from the theoretical formalism, just as Einstein's postulates allow the Lorentz contraction to emerge as a natural consequence.

2.7 Ernst Mach: visionary/reactionary

I digress slightly here to discuss Ernst Mach, a prominent figure in nineteenth-century physics, because he probably exemplifies more than anyone else the irony discussed in this chapter. He exemplified, on the one hand, the virtue of humble submission and obedience to nature's empirical messages; and on the other hand, the philosophical mistake of assuming that those empirical phenomena are all there is, or that knowledge cannot, or should not, go beyond them. As a strict empiricist, Mach insisted that all knowledge is based on sensation or observation – a position which of course confines any empiricist to knowledge about the world of appearance only. Yet it does not follow that *the only thing that exists* is appearances, as noted earlier; and here I endorse von Weizsaecker's dictum that "What is observed certainly exists; about what is not observed we are still free to make suitable assumptions. We use that freedom to avoid paradoxes."[22] (Descartes has more pungent remarks for the strict empiricist, as we will see shortly.)

Thus, while I agree with Mach's eliminativist[23] account of spacetime as fundamentally based on comparisons (i.e., I adopt a relational view of spacetime), that does not mean that the interpretation of all physical theories which were discovered through the application of mathematical analysis to observations must be limited to subjective sensations, as Mach unnecessarily (and I believe mistakenly) concludes in the last clause below:

> . . . we do not measure mere space; we require a material standard of measurement, and with this the whole system of manifold sensations is brought back again. It is only intuitional sense-presentations that can lead to the formulation of the equations of physics, and *it is precisely in such presentations that the interpretation of these equations consists* . . . [*AS*: 343] (emphasis added)

Thus, Mach's justified insistence that theory *construction* be grounded in observation slides unjustifiably into categorical antirealism about possible unobservable entities pointed to by those theories. As noted previously, this is a logical and methodological error, unambiguously revealed as such when Mach's refusal to entertain the existence of atoms – because they were unobservable – was shown to have been on the wrong side of scientific progress. One can acknowledge that perhaps what we think of as "spacetime" can be reduced to an account of the ordering of sensations (also known as material objects), but it does not necessarily follow that there is *nothing more to reality* than sensations. The ordering we discover can be seen as an objective property of reality insofar as all our observations conform to it and it cannot be altered by purely subjective means (i.e., by imagining or desiring it to be different). Thus, objective reality may be something real, even if not directly observable, which is capable of giving rise to sensations (i.e., observations, or actualized events).

The unjustified assumption that because our knowledge of reality is derived largely from sensation, our interpretation of theories and our understanding of reality must be *limited* to accounts of sensation, is subjected to rather harsh criticism by Descartes in his *Treatise on Light*. I quote generously here, as Descartes takes a while to establish his point (italics added for emphasis):

[22] Private communication, first quoted in Cramer (1986).
[23] A term meaning that the concept under study does not correspond to an independently existing entity or substance.

> . . . the spaces where we sense nothing are filled with the same matter, and contain at least as much of that matter, as those occupied by the bodies that we sense. Thus, for example, when a vessel is full of gold or lead, it nonetheless contains no more matter than when we think it is empty. This may well seem strange to many *whose [powers of] reasoning do not extend beyond their fingertips and who think there is nothing in the world except what they touch*. But when you have considered for a bit what makes us sense a body or not sense it, I am sure you will find nothing incredible in the above. For you will know clearly that, far from all the things around us being sensible, it is on the contrary those that are there most of the time that can be sensed the least, and those that are always there that can never be sensed at all.
>
> The heat of our heart is quite great, but we do not feel it because it is always there. The weight of our body is not small, but it does not discomfort us. We do not even feel the weight of our clothes because we are accustomed to wearing them. The reason for this is clear enough; for it is certain that we cannot sense any body unless *it is the cause of some change* in our sensory organs, i.e. unless it moves in some way the small parts of the matter of which those organs are composed. The objects that are not always present can well do this, provided only that they have force enough; for, if they corrupt something there while they act, that can be repaired afterward by nature, when they are no longer acting. But if those that continually touch us ever had the power to produce any change in our senses, and to move any parts of their matter, in order to move them they had perforce to separate them entirely from the others at the beginning of our life, and thus they can have left there only those that completely resist their action and by means of which they cannot be sensed in any way. Whence you see that it is no wonder that there are many spaces about us in which we sense no body, even though they contain bodies no less than those in which we sense them the most. (*Treatise on Light*, chapter 4; italics added)

Thus (in admittedly uncharitable language), Descartes argues that it is a mistake to assume that nothing exists beyond what we sense, as our material senses can only detect *change*, not entities that are always present or that are incapable of activating our sense organs. It is widely supposed that Descartes' metaphysics, which postulated a dynamic plenum rather than a void underlying observable matter, was a quaint piece of "moribund metaphysics" (to use van Fraassen's term)[24] that was largely discredited by Newton's theories. Yet, arguably, Descartes can now be seen as having presaged the development of relativistic quantum theory, which has taught us that what Newton thought of as the "void" is far from empty.[25] So we would do well to reacquaint ourselves with Descartes' views on scientific methodology. We should also consider the von Weizsaecker quote above that "what is observed certainly exists; about what is not observed we are still free to make suitable

[24] For example, van Fraassen (2004, p. 3).

[25] For example, the latest cosmological studies suggest that the vacuum contains "dark energy," which affects the expansion of the universe; also, the Standard Model of particle theory postulates a background field, the Higgs field, which is responsible for the finite masses of particles. There is also the basic zero-point energy of the vacuum, a quantum-mechanical effect. This energy can have physically measurable effects; for example, a detector in uniform acceleration through the "vacuum" will detect thermal radiation (the Unruh effect). There is continual particle/antiparticle creation arising from the vacuum. Overall, an astonishing amount of activity goes on in so-called "empty space."

assumptions." Such a "suitable assumption," as remarked earlier, was the existence of atoms. So despite Mach's insights into the importance of recognizing how our knowledge is obtained largely through sensation, he refused to countenance a crucial theoretical construct – the atom – which led to important scientific breakthroughs. The lesson, I suggest, is to acknowledge that we should not let metaphysical preconceptions get in the way of observations and theory construction based on those observations, but we should *not* uncritically assume from the success of that approach that, as Descartes says, "there is nothing in the world except what [we] touch."

2.8 Quantum theory and the noumenal realm

So what can be gained by exploring the possibility that certain aspects of the quantum formalism typically thought to have only operational significance (e.g., dual states or bracs, denoted as $\langle\Psi|$) may indeed have ontological significance? Recall Zeilinger's observation that the "individual event" remains resistant to causal description, along with similar observations by the founders of quantum theory. For example, according to Jammer (1993), Bohr referred to such events, such as the inherently unpredictable transitions of electrons in atoms from one stationary state to another,[26] as "transcending the frame of space and time."[27] As discussed earlier in this chapter, Bohr regarded spacetime concepts (indeed, all "classical" concepts) as prerequisites for the endeavor of gaining physical knowledge of the world; thus he explicitly restricted what counted as legitimate knowledge to that of the world of appearance, in Kantian terms. Yet the significance of his quoted remark is that it clearly implies *there are real physical events which transcend the boundaries of the observable universe*. For surely Bohr has to acknowledge that stationary states were instantiated in nature, and that transitions between them did occur, as this much is empirically corroborated.

Recall that Bohr insisted that physics concerns "what we can say about nature." But what is the "we" in this context? Is it ordinary language? Or is it the mathematical language of our best theories? If the former, obviously it is very difficult, if not impossible, to talk about events which "transcend the frame of space and time." But even Bohr implicitly admitted, as noted above, that such events occur. Indeed, the very theory he helped invent is what led him to make this observation. Does that not, then, mean that the formal aspects of physical theory *can* point to heretofore unknown aspects of physical reality, however difficult it might be to talk about

[26] Stationary states are states whose wave functions do not change with time. An atom's discrete energy levels correspond to such states.
[27] As quoted in Jammer (1993, p. 189).

them – that physics *can* be more than what we can "say about Nature" in ordinary, classically *anschaulich* terms?

I believe that the answer to that question is "yes." The fact that quantum theory, in Bohr's words, seems to point to entities and/or processes transcending the frame of space and time means that quantum theory can reasonably be thought of as (at least in part) a theory about the noumenal realm.[28] That is, since concepts like space and time are considered vital for gaining and communicating knowledge about the world of appearance, processes that "transcend" those concepts must be processes belonging to the noumenal realm, which transcends the world of appearance. Therefore, I claim that the truly revolutionary message of quantum theory is *not* that we should stop asking questions about the nature of reality; on the contrary, the message is that quantum theory is offering a new and strange kind of answer about an aspect of reality traditionally pronounced "off limits" by Kant and those (like Bohr) who subscribe to the notion that physical theory can only be about the world of appearance. That this methodological restriction should be abandoned is supported by Bohr's own comment about certain quantum processes "transcending space and time," which, contrary to his other pronouncements, unambiguously testifies to knowledge gained from quantum theory concerning the possible existence of a realm transcending space and time.

Indeed, as Einstein and others have noted, there appears to be a deep and significant connection between certain mathematical objects and physical reality – were that not the case, the whole field of theoretical physics would be without power or purpose in providing an account of the empirical realm. There is ample precedent for entities and procedures that seem purely formal and abstract turning out to have concrete physical relevance. For example, in the words of Freeman Dyson, the mathematicians of the nineteenth century "had discovered that the theory of functions became far deeper and more powerful when it was extended from real to complex numbers. But they always thought of complex numbers as an artificial construction, invented by human mathematicians as a useful and elegant abstraction from real life. It never entered their heads that this artificial number system that they had invented was in fact the ground on which atoms move. They never imagined that nature had got there first" (Dyson, 2009).

2.9 Science as the endeavor to understand reality

As argued in the foregoing, I believe that quantum theory can present us with a new kind of understanding of nature, based on a wholly new kind of model, *if* we listen carefully and open-mindedly to what the formalism is saying. I take such a new

[28] More precisely, that the domain of quantum theory includes the noumenal realm as a component.

understanding of nature afforded by a theory as an "explanation" of the empirical phenomena in the domain of the theory. However, for those who demand that a model be constructed out of actual, "things of the facts" (by this I mean ordinary, causal, "classical" facts as referred to by the oft-used term "local realism"), there can of course be no such "explanation," as nearly a century of determined attempts has revealed. The failure of classical model-making has been well-established and has largely been answered by a turn to strict empiricism and even frank instrumentalism by many researchers who assume, with Bohr, that all models must be classical. Empiricist approaches are essentially Bohrian in character, denying that the job of science is to "understand how nature is" and rejecting the whole idea of model construction as a misguided "demand for explanation" that need not be met (cf. van Fraassen, 1991, p. 372).[29] In this perspective, it is seen as virtuous to renounce explanation in science, and a sign of enlightened wisdom to content ourselves with classifying and predicting phenomena. But, as argued above, this position does not follow logically from the failure of inappropriate mechanical, deterministic, local (classical) models, and it is at odds with arguably the most important and exciting aspect of the scientific mission: the *discovery* of previously unseen and unknown aspects of reality (a case in point being the atom and its constituents). If we reconceptualize the process of modeling in light of quantum theory, perhaps we can find a new and more fruitful means of discovery.[30]

[29] Moreover, van Fraassen (1991, p. 24) conflates the possible existence of "randomness" with "no explanation" in passages such as this, addressing specific outcomes or asymmetries with no apparent antecedent cause: "[Pierre] Curie's putative principle [that 'an asymmetry can only come from an asymmetry'] betokens only a thirst for hidden variables, for hidden structure that will explain, will answer *why*? – and nature may simply reject the question." In this regard, there may not be a causal, determinate, mechanistic account, but that doesn't mean that there can be *no* account of relevant and interesting additional structure, so the pursuing of such an account is not merely evidence of a futile 'thirst for hidden variables.' For instance, there is no *deterministic* account of how one ground state is selected from among many possible ones in spontaneous symmetry breaking, yet one can certainly give an account of the process of symmetry breaking in terms of an additional structure which sets the stage for the circumstance of symmetry breaking. This point is addressed in Chapter 4.

[30] It should also be noted in this context that the recent theorem of Pusey *et al.* (2011) appears to rule out instrumentalist approaches based on taking the quantum state as characterizing observer knowledge only, and subject to instantiation by more than one (hidden) ontological state. If Bohr's interpretation of quantum theory assumes precisely this kind of epistemic interpretation of quantum states, his views are then refuted by the Pusey theorem, and most of the qualitative arguments in this chapter are unnecessary (even if still valid). But that remains a topic for future research. (It could be argued that Bohr simply ruled out the existence of hidden ontological states, so that the conditions of proving the Pusey theorem would not be available. After all, he did assert that "there is no quantum world.")

3

The original TI: fundamentals

3.1 Background

The transactional interpretation of quantum mechanics was first proposed by John G. Cramer in a series of papers in the 1980s (Cramer, 1980, 1983, 1986, 1988). The 1986 paper presented the key ideas and showed how the interpretation gives rise to a physical basis for the Born Rule which prescribes that the probability of an event is given by the square of the wave function corresponding to that event. TI was originally inspired by the Wheeler–Feynman (WF) time-symmetric theory of classical electrodynamics (Wheeler and Feynman, 1945, 1949). The WF theory proposed that radiation is a time-symmetric process, in which a charge emits a field in the form of half-retarded, half-advanced solutions to the wave equation, and the response of absorbers combines with that primary field to create a radiative process that transfers energy from an emitter to an absorber. This process is symbolized by a "handshake." Let's first review the WF proposal, and then we'll see how TI generalizes the idea to the quantum domain.

3.1.1 The wave equation

The wave equation for any field relates the spatial variation of the field to its time variation. For a generic massless wave field denoted by Φ, the wave equation in the absence of sources (called the "homogeneous wave equation") has the form

$$\nabla^2\Phi - \frac{1}{v^2}\frac{\partial^2\Phi}{\partial t^2} = 0 \quad (3.1)$$

where v is the speed of propagation of the wave.

To take into account a specific source for the field, a "current" J is added to the right-hand side, giving the "inhomogeneous wave equation"

$$\nabla^2\Phi - \frac{1}{v^2}\frac{\partial^2\Phi}{\partial t^2} = J \tag{3.2}$$

A current J can be a point source such as an electron, or a more extended object (charge distribution) capable of coupling to the electromagnetic field. "Coupling" means having the ability or tendency to emit or absorb photons, the quanta of electromagnetic radiation; the ability for a current to couple in this way to the electromagnetic field is indicated by saying that an object has charge.[1]

3.1.2 Coupling and absorption in TI

I digress briefly here to note that the concept of coupling is important for understanding the process of absorption in TI, which is often misunderstood. Under TI, an "absorber" is an entity which generates confirmation waves (CW) in response to an emitted offer wave (OW). (Both these concepts – OW and CW – are defined in more technical terms in Section 3.2 below.) The generation of a CW needs to be carefully distinguished from "absorption" meaning simply the absorption of energy, since not all absorbers will in fact receive the energy from a given emitter. In general, there will be several or many absorbers sending CW back to an emitter, but only one of them can receive the emitted energy. This is purely a quantum effect, since the original classical WF absorber theory treats energy as a continuous quantity that is distributed to all responding absorbers. It is the quantum level that creates a semantic difficulty in that there are entities (absorbers) that *participate* in the absorption process by generating CW, but don't necessarily end up receiving energy. In everyday terms, these are like sweepstake entrants that are necessary for the game to be played, but who do not win it.

A longstanding objection to the TI picture has been that the circumstances surrounding absorption are not well-defined, and that "absorber" is a primitive term. The objection argues that this makes the TI account dependent on arbitrarily decreeing a measurement "completed" based on pragmatic considerations, tenuous "irreversibility" arguments, the need to express results in "classical terms," or on the consciousness of an observer, as in traditional approaches which remain subject to the measurement problem (recall Chapter 1). This objection is addressed and resolved in the current approach as follows. TI can indeed provide a non-arbitrary (though not deterministic) account for the circumstances surrounding absorption in terms of coupling between fields. Since this is a relativistic concept, I defer those details to Chapter 6; but in a nutshell, I propose that "absorption" in TI simply means

[1] More precisely, "coupling" means that a current has a non-zero amplitude to emit or absorb a photon.

annihilation of a quantum state, which is a perfectly well-defined physical process in the relativistic domain. The fact that objections to TI can be resolved at the relativistic level underscores both (1) the ability of TI to accommodate relativity and (2) the necessity to include the relativistic domain to resolve the measurement problem (as is also addressed in Chapter 6).

3.1.3 Solutions of the wave equation

Returning now to consider the wave equation and its possible solutions, we first need to review some basic features of waves. Any generic wave has a wavelength λ and a frequency f; the speed v of the wave is simply their product

$$v = \lambda f \tag{3.3}$$

It is customary, for notational convenience, to express λ and f in terms of a "wave number" k and an angular frequency ω, respectively, where

$$k = \frac{2\pi}{\lambda} \text{ and} \tag{3.4a}$$

$$\omega = 2\pi f \tag{3.4b}$$

Thus, the propagation speed of the wave can also be written

$$v = \lambda f = \frac{2\pi}{k}\frac{\omega}{2\pi} = \frac{\omega}{k} \tag{3.5}$$

The above is termed the "phase velocity"; it specifies the distance traveled by a particular wave crest in unit time (see Figure 3.1).

In the empirical world, we always seem to see waves diverging outward into space from the past to the future (i.e., from earlier times to later times), as shown in Figure 3.2.[2]

This type of wave propagation is called "retarded" propagation, and corresponds to a solution to (3.1) of the form[3]

[2] We should not, however, *equate* the divergence of the wave with the fact that it is a retarded solution. Retarded waves are simply waves that are created at a source and later encounter an absorber. Such waves could be in a light pipe or transmission line and do not necessarily show spherical wave divergence.

[3] For simplicity, I neglect constant coefficients. In addition, this presentation is a heuristic one in a single spatial dimension x, so it does not reflect the distinction between solutions to the homogeneous and inhomogeneous equations. Strictly speaking, "advanced" and "retarded" solutions only apply to the inhomogeneous wave equation (i.e., the equation with sources) in three spatial dimensions.

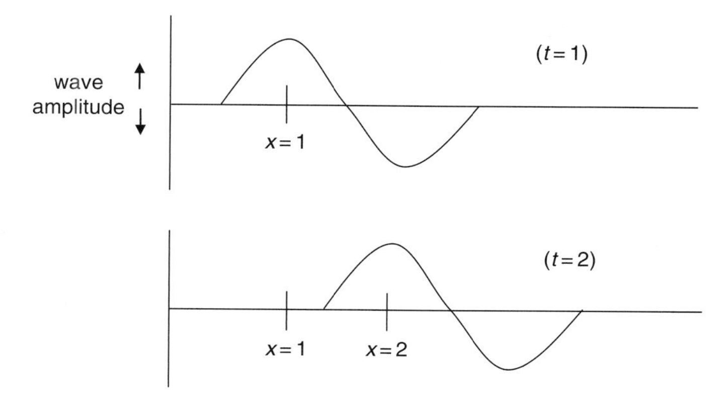

Figure 3.1 The wave crest depicted here travels from $x = 1$ (m) to $x = 2$ (m) in unit time (s), so this wave's phase velocity is 1 m/s. Only one wavelength is shown for simplicity.

Figure 3.2 A falling raindrop creates diverging ripples on the surface of a pond. *Source*: Salvatore Vuono/FreeDigitalPhotos.net

$$\Phi_r(x, t) = \exp\left[\frac{i}{\hbar}(kx - \omega t)\right] \tag{3.6}$$

We can understand the solution of (3.6) as propagating into the future by seeing that the value of x for any point of constant phase (such as a wave crest referred to above) increases with increasing time. For example, when $kx = \omega t$, we have $\phi = 0$. In order to keep the phase constant in this expression as t increases, x must increase; so the wave propagates spatially in the same direction as its temporal propagation (see Figure 3.1).

However, an equally (mathematically) valid solution exists in the form of "advanced" propagation:

$$\Phi_a(x,t) = \exp\left[\frac{i}{\hbar}(kx + \omega t)\right] \tag{3.7}$$

Let's examine the behavior of the phase of this solution as we did for the retarded case. As t increases, the value of the spatial index x must *decrease* to keep the phase constant (i.e., to keep track of the same spot on the wave such as a crest or trough). For the spatial index to increase, the temporal index t must *decrease* (i.e., the wave must propagate "into the past"). If we consider the more realistic 3-dimensional situation in which a point source gives rise to the field solutions under consideration, the spatial coordinate x changes to r, which tells us the radial distance from an emitting source.[4]

The retarded solution corresponds to a set of spherical wave fronts (sets of spatial points of constant phase) that diverge with increasing t; i.e., r increases with increasing t. In contrast, the advanced solution corresponds to cases in which the spatial and temporal indices increment in opposing directions. This give us either (1) a set of wave fronts converging onto the source from the past, or (2) a set of wave fronts diverging *from* the source *into* the past (depending on which way we choose to orient the "flow" of events with respect to a spacetime diagram). These 3-dimensional forms are illustrated in Figure 3.3.

In Figure 3.3, we can think of someone with a stopwatch standing at the origin of a coordinate system and shining a flashlight for a split second when their stopwatch says $t = 0$. As we map the person's experience on a spacetime diagram (consider (a) first), he is at the center of an ever-widening sphere of concentric wave fronts. Because we can't represent three spatial directions plus a time direction on paper, these spherical wave fronts have to be pictured as a series of widening circles (we neglect one spatial dimension so the spheres get flattened to circles). The person's worldline (not pictured) is a straight vertical line through the center of all the circles.

Figure 3.3(a) is the usual "retarded" wave that diverges as the time index increases. In contrast, Figure 3.3(b) shows the "advanced" wave which diverges as the time index *decreases* – that is, it *propagates into the past*. If we observe the advanced wave from our usual temporal orientation – i.e., moving "forward" in time – the advanced wave appears to emerge from all directions and to converge onto the source.

[4] In addition, there is a factor of $1/r$, assuming spherical symmetry.

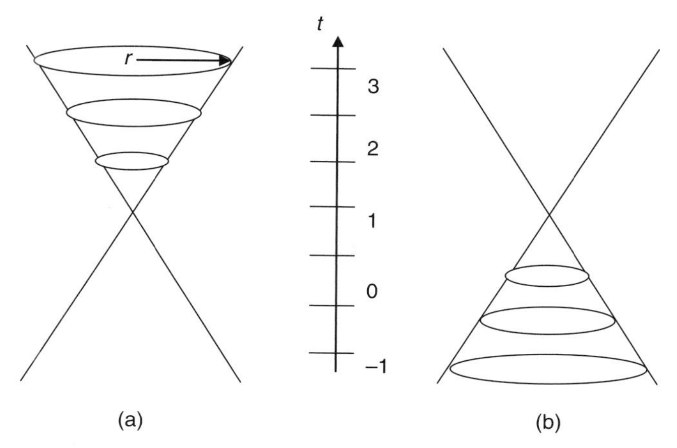

Figure 3.3 Pictured in 2(space)+1(time) dimensions are (a) the retarded wave solution; (b) the advanced wave solution, both with respect to a hypothetical source at the origin (where the lightlike diagonal lines cross). The foreshortened circles are actually spherical wave fronts in 3+1 dimensions.

3.1.4 The Wheeler–Feynman theory

The Wheeler–Feynman proposal is that all radiation sources emit half their radiation as retarded and half as advanced; this solution is termed a "time-symmetric" solution. So, in terms of Figure 3.3, the source at the origin (where the lightlike diagonals cross) emits equal amounts of both (a) and (b). Other charges respond to the emitted time-symmetric field by emitting their own symmetric field, but exactly out of phase with the stimulating field. Using the ability of radiation fields to add, Wheeler and Feynman show that, if the universe is a "light-tight box,"[5] the overall advanced response ("echo") of all absorbers to the retarded radiation from any particular emitter amounts to precisely the same field as that original half-strength retarded radiation field from the emitter.

The above process results in two distinct effects. (1) The two fields add; thus, from the point of view of an observer, the retarded field from the source appears to be full strength and the advanced components cancel. (2) In addition, the absorber response provides for a "free field" component[6] that must be assumed in an ad hoc manner in the standard theory (which assumes that the source emits only a retarded field) in order to account for the loss of energy by a radiating charge. This twofold process, wherein the advanced field from the absorber (1) superimposes constructively with the retarded field of an emitter and (2) provides for energy transfer from

[5] This means that any emitted radiation is fully absorbed; no retarded radiation escapes to future infinity.
[6] The "free field" is the difference of the retarded and advanced solutions. It has the properties of a field that does not arise from (or converge onto) a source (or sink), but simply exists independently.

the emitter to the absorber, forms the basis for the "transaction" in TI. The second aspect is what allows TI to say that a current exists between an emitter and absorber which can transfer energy, where that current is of the form $\Psi^*\Psi$, reflecting the Born Rule (this is discussed in more detail in Section 3.2).

As Price has noted (1996), the advanced wave could be identified with an absorption process. That is, one can imagine the advanced wave pictured in Figure 3.3(b) as converging onto a radiation "sink" instead of onto an emitter (imagine water from all areas of a basin converging as it goes down the drain). He argues that the WF approach of having both emitters and absorbers radiate half-retarded and half-advanced is a kind of "symmetry overkill" (my term) and that all one needs for symmetry with respect to radiative processes is to associate emitters with 100% retarded radiation and absorbers with 100% advanced radiation. Then, the problem of explaining the apparent predominance of retarded radiation becomes one of explaining why we see large-scale emitters (such as stars) but not large-scale sinks of radiation.

However, one basis for disagreement with Price's conclusion that the above restores symmetry to radiative processes is that it treats propagation with respect to time differently from propagation with respect to space. That is, radiative processes – both emission and absorption – are *isotropic* with respect to space, meaning that they involve propagation in all spatial directions (which is why we have spherical wave fronts as opposed to semi-spherical ones). In other words, radiation is emitted not just in the positive (or negative) x, y and z directions, but in both. Since the breakthroughs of special relativity are inextricably bound to the intermingling of space and time,[7] a more relativistically consistent approach is to allow that radiation is also symmetric (the 1-dimensional equivalent of isotropic) with respect to time. Thus, radiation should be emitted in the positive and negative t directions as well. Then the observed asymmetry of radiation is related to the boundary conditions of the universe, just as the direction of the flow of heat in a thermodynamic system is related to its boundary conditions. I return to this issue in Chapter 8.

As noted in Cramer (1986), the Wheeler–Feynman approach to dealing with classical radiation theory fell out of favor because it assumed that a radiation source could not interact with its own field; but this "self-interaction" or "self-energy" was found to be necessary, at least at the quantum level, for certain known empirical effects such as the Lamb shift (cf. Berestetskii *et al.*, 2004, p. 535). Also, the explicit dependence of field configurations on future states as well as past states made

[7] For example, the Lorentz transformations relating spatial distances x and x' in two different frames O and O' must introduce a *time* coordinate from one or the other frame. Space and time are routinely mixed in basic relativistic calculations. In fact, the Lorentz transformations can be derived from imagining a light pulse expanding in a 4-dimensional hypersphere: the time index is squared just as the spatial indices are.

computations more difficult than in the standard approach. Feynman's subsequent work on the Lagrangian (or action-based) approach to quantum mechanics and quantum field theory, rather than the usual Hamiltonian (or energy-based) approach, was undertaken at least in part to see if these difficulties could be surmounted. As it turns out, the Lagrangian approach can be seen as empirically equivalent to the Hamiltonian approach (cf. Zee, 2010, pp. 61–3). Most researchers prefer using the Hamiltonian approach because calculations are performed more easily in most cases with the latter, but it may be argued that the Lagrangian formulation is theoretically more fundamental since it is relativistically covariant. This is because the Hamiltonian approach singles out energy as a privileged quantity, and since energy is conjugate to time, it also implicitly designates a particular time coordinate as privileged, which is inconsistent with relativistic covariance.

Furthermore, Davies (1970, 1971, 1972) extended the basic Wheeler–Feynman approach to quantum electrodynamics, which included the possibility of self-interaction based on the indistinguishability of currents (i.e., the quantum feature that, for example, all electrons are indistinguishable aside from measurable properties such as momentum or spin). The basic conclusion is that there is nothing theoretically wrong with the Wheeler–Feynman approach (in fact, in many ways it is theoretically superior due to its more symmetrical and covariant features as discussed above); it merely has not been as suitable for practical calculations.

Feynman's motivation for his direct action approach (the Wheeler–Feynman absorber theory) had been to eliminate the electromagnetic field as an independent entity, and he later decided that this could not be done because of the need for self-energy. TI reconciles the apparent tension between a "direct action" theory and self-energy by accounting for the self-energy in transactional terms.[8] In fact, the self-energy interaction has a particularly natural interpretation under TI: it is simply a case in which an emitter acts as an absorber for its own offer wave.

3.2 Basic concepts of TI

Cramer (1986) specified the ways in which TI differed from traditional approaches to interpreting quantum mechanics (in particular the Copenhagen interpretation) and argued instead for a straightforward realist approach in which the theoretical quantum state $|\Psi\rangle$ and its adjoint $\langle\Psi|$ represent real physical entities in a time-symmetric interpretation based on the basic Wheeler–Feynman formalism. It showed that the Born Rule for calculating the predicted probabilities of observable

[8] Of course, there remains the issue of theoretically infinite self-energy. As observed in Chapter 6, this likely stems from the fact that it is not physically legitimate to take the infinite mathematical limit of the perturbation expansion.

events arises naturally from the interaction of offer waves (represented mathematically by the usual quantum state or wave function) and confirmation waves (represented by the adjoint quantum state or complex conjugate wave function). In the remainder of this chapter I review the key features of this original proposal.

3.2.1 Emitters and absorbers

In simple terms, emitters and absorbers are those entities in standard physics that can emit or absorb another quantum. More technically, as addressed briefly in Section 3.1.2, emitters and absorbers are field currents that can couple to other fields, which means that they have an amplitude to emit or absorb quanta of a field. Emission or absorption occurs when creation or annihilation of a quantum state (respectively) takes place. Examples are electrons that can emit and absorb photons: an electron can serve as an emitter or as an absorber of photons. A macroscopic emitter of electrons could be a piece of heated metal in which the conduction electrons are excited to the point where they become liberated from the surface of the metal (this is called "thermionic emission"). A macroscopic absorber of electrons is any substance whose molecules' potential energy can be lowered by binding with an electron. As noted in Section 3.1.2, an absorber is an entity that responds to an emitted offer wave with a confirmation wave, whether or not that particular entity actually ends up receiving the energy.

3.2.2 Offer waves and confirmation waves

The term "offer wave" (OW) denotes the entity referred to by the usual quantum state $|\Psi\rangle$, which corresponds to the retarded component of the field in the Wheeler–Feynman account. An OW is what is emitted by an emitter (along with the emitter's advanced wave component). A "confirmation wave" (CW) is the advanced component of the response field generated by an absorber and is represented by the dual state or "brac" $\langle\Phi|$ (the state labels are arbitrary here). The CW corresponds to the advanced component of the field in the Wheeler–Feynman account. The process whereby the absorber's advanced field (CW) reinforces the emitter's retarded field, and the remaining advanced component from the emitter and retarded component from the absorber are cancelled, is illustrated in Figure 3.4.

Since it is well known that the operation of time reversal takes "kets" into "bracs,"[9] this gives a natural time-symmetric interpretation of the ubiquitous inner product quantities appearing in quantum theory, such as $\langle\Phi|\Psi\rangle$. We come across an inner product form when taking into account the fact that an absorber corresponding to the

[9] See, e.g., Sakurai (1984, pp. 273–4).

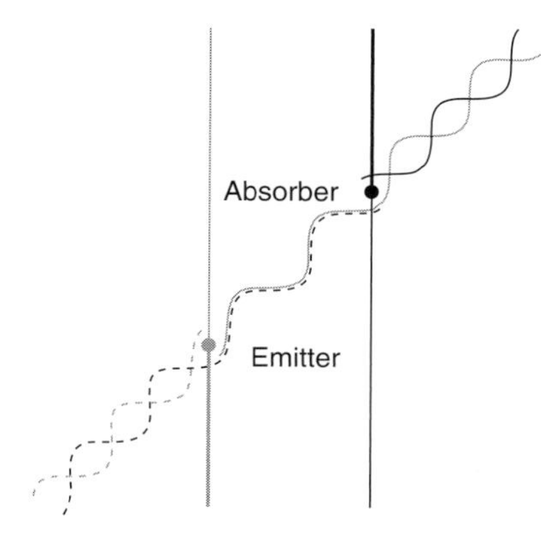

Figure 3.4 Here the absorber represents all the microscopic absorbers in a "light tight box." The advanced field (dashed line) from the absorber exactly reinforces the retarded field (solid line) between the emitter and absorber and exactly cancels the advanced field from the emitter and the retarded field from the absorber, so all that remains is a fully retarded wave carrying energy from the emitter to the absorber.

property labeled Φ can only absorb (annihilate) that component of any OW encountering it. This can be thought of as "attenuation" of the original OW from the perspective of the absorber corresponding to Φ: the component of the OW in the state labeled Ψ reaching an absorber corresponding to the state labeled Φ will be the projection of $|\Psi\rangle$ onto $|\Phi\rangle$, or $\langle\Phi|\Psi\rangle|\Phi\rangle$.

In the next section I review how the Born Rule, or the probability of an outcome corresponding to the property Φ for a system prepared in state $|\Psi\rangle$: $P(\Phi|\Psi) = |\langle\Phi|\Psi\rangle|^2$, arises naturally in TI.

3.2.3 *The Born Rule is revealed in TI*

Let us consider the more general case in which an emitted OW labeled $|S\rangle$, from a source S of quanta (such as a laser), encounters absorbers labeled by properties A, B, C, D, ... (see Figure 3.5). As described in the previous section, the component absorbed by A is $\langle a|s\rangle|a\rangle$ and the component absorbed by B is $\langle b|s\rangle|b\rangle$, etc. Each absorption results in the advanced CW $\langle s|a\rangle\langle a|$ and $\langle s|b\rangle\langle b|$, etc. (this is the "response of the absorber" to the emitter). The product of the OW and CW amplitudes gives the Born Rule for the probability of the outcome, e.g., $P(A|S) = \langle s|a\rangle\langle a|s\rangle = |\langle a|s\rangle|^2$.

The preceding account leads to a weighted set of "competing" possible transactions that we can call "incipient transactions." Note that all possible transactions are associated with projection operators; i.e., matching "final" OW and CW

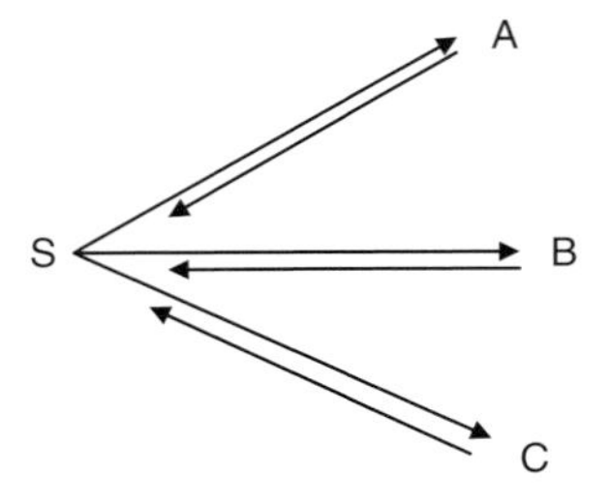

Figure 3.5 An offer wave $|S\rangle$ can be resolved into various components corresponding to the properties of absorbers A, B, C, D . . . The product of a particular OW component $\langle a|s\rangle|a\rangle$ with its corresponding CW component $\langle s|a\rangle\langle a|$ reflects the Born Rule which tells us that the probability of the result corresponding to the projection operator $|a\rangle\langle a|$ is equal to $\langle a|s\rangle\langle s|a\rangle = |\langle a|s\rangle|^2$.

components. Thus, the weighted set of incipient transactions is just von Neumann's "Process 1," discussed in Chapter 1.[10] To review, von Neumann proposed that, upon measurement of an observable O with possible values X_i (these correspond to A, B, C, D above) on a system prepared in state $|\Psi\rangle$, the system's state undergoes a change from a "pure state" to a "mixed state," i.e.,

$$|\Psi\rangle\langle\Psi|\rightarrow\sum_i|\langle\Psi|X_i\rangle|^2|X_i\rangle\langle X_i|. \tag{3.8}$$

However, the notorious problem with the von Neumann formulation was that there seemed to be no way to determine when, why, or how the pure state should undergo such a transformation. If we take into account the physical process of absorption (i.e., state annihilation), "Process 1" becomes completely non-mysterious. It is just the process whereby the CW are returned to the emitter from all absorbers capable of responding, and a set of incipient transactions is established.

The "mixed state" on the right-hand side of (3.8) represents a set of incipient transactions, of which (in general) only one can be actualized. However, the presence of absorbers defines unambiguously the basis with respect to which the offer wave must be decomposed, thus eliminating many of the perplexing ambiguities often present in discussions of the quantum state (which can *theoretically* be expressed in myriad such bases). Here, the "observable" being measured is the operator defined by the sum of the incipient transactions represented by $|X_i\rangle\langle X_i|$ in (3.8), where each is multiplied by its associated eigenvalue (i.e., the "value of the observable" corresponding to that outcome). The latter is referred to in the literature as the "spectral decomposition" of the observable.

[10] See also Bub (1997, p. 34).

The weighted set of incipient transactions corresponds to a classical probability space in which the weights can be straightforwardly interpreted as the probability that the answer "yes" can be consistently applied to questions such as "is the system in state X_k?" There is true collapse in TI, in that the property ultimately selected is stochastically actualized with the corresponding probability. This collapse is understood as a type of symmetry breaking; the latter is discussed in detail in Chapter 4.

3.3 "Measurement" is well-defined in TI

As is evident in the foregoing, the key advantage of TI over other "collapse"-type interpretations is that the notion of "measurement" is unambiguously defined in physical terms, without appeal to the "consciousness" of an external observer. In this section, I compare TI's treatment of measurement with competing accounts and see how it provides a solution to the measurement problem, in the sense of making clear at what point a measurement can be said to actually occur.

3.3.1 TI's advantages over traditional "collapse" interpretations

A system undergoing measurement in TI is actualized in a particular state, corresponding to the actualization of a particular event/property in classical terms. The account is not relational in that this event definitely occurs; it is not "contextual" or defined only relative to any external system or observer.

How is this achieved? Very simply, by taking into account that *absorption is a real physical process*. This is certainly the case in relativistic quantum field theories: one cannot arrive at a correct empirical prediction without taking absorption into account. Indeed, absorption (i.e., annihilation) is a key element of the definition of the field operators used in any calculation of probabilities of empirical events. Such calculations routinely involve taking expectation values in which quanta are created and quanta are destroyed. If such calculations refer to anything physical (the basic realist assumption), both processes are physical processes. However, for the past century or so, interpretations of non-relativistic quantum theory have completely disregarded the absorption process, granting physicality only to emission processes giving rise to quantum objects that are described by the usual (retarded) quantum states ("kets"). They have thus considered non-relativistic quantum mechanics – which is just a limiting case of quantum theory – only in a particular form (as applying only to emission) and in isolation from its relativistic application; and this, I suggest, is what has prevented the ability of such interpretations to account for measurement in physical terms.

Specifically, a measurement or determinate event (i.e., it does not have to be a formal "measurement" conducted by an observer) occurs whenever annihilation of one or more free quanta occurs.[11] In terms of relativistic quantum theory, absorption corresponds to the action of annihilation operators on free quanta, just as emission corresponds to the action of creation operators on the vacuum state.[12]

As noted earlier, a common objection to TI is the claim that absorbers are not well-defined, but this objection apparently ignores the fact that absorbers *are* well-defined objects throughout physics. If emitters are taken as well-defined – that is, if we can assert that it makes ontological sense to say that the entity described by a quantum state is emitted (created) – then one cannot consistently argue that it doesn't make ontological sense to say that the entity described by a quantum state is absorbed (annihilated).

As noted in Section 3.1.2, one source of confusion surrounding TI is that "absorption" is sometimes conflated with "detection" – that is, with empirically detectable transfer of energy from an emitter to an absorber. But (in terms of quantum field theory) an annihilation operator corresponding to property A can act without necessarily resulting in an actualized event A, just as the creation operator corresponding to property B can act without resulting in an actualized event corresponding to property B. For example, the ket $|p\rangle$ can be written in terms of a creation operator as $a^{\dagger}(p)|0\rangle$, which can be understood as the creation of the *possibility* of property p from the vacuum. Now, recall that a particular momentum state can be written as an infinite sum of all possible position states $|x\rangle$.[13] If a measurement of position is then performed, the quantum will be detected at some position x. What happened to the property p? It was not actualized. Note that the brac $\langle p|$ can be written as $\langle 0|a(p)$. This corresponds to the destruction of *the possibility* of property p, just as $|p\rangle$ corresponds to the creation of the possibility of property p.

Alternatively, one can create a quantum state (offer wave) corresponding to spin "up along x," $|x = \text{up}\rangle$, and then allow it to interact with a Stern–Gerlach device oriented along z. Absorbers placed at each of the outputs corresponding to "up" and "down" along z both act on that state to destroy (absorb) the corresponding property, but a particle is only actually detected by one of them (i.e., conserved physical quantities are only transferred to one of the absorbers). The key point is that *the absorption (annihilation) of the entity described by a quantum state is not the same as empirical detection of an actual quantum*. The identification of quantum states as

[11] I add the qualifier "free particles" because annihilations associated with *virtual* particles have an amplitude less than unity of being accompanied by confirmation waves. This topic is discussed in Chapter 6.

[12] For example, the action of the creation operator for momentum p on the vacuum state $|0\rangle$ yields the state $|p\rangle$ ($|p\rangle$ is emitted), while the action of the annihilation operator for momentum p on the state $|p\rangle$ yields the vacuum state ($|p\rangle$ is absorbed).

[13] Strictly speaking, there are field states $\phi(x)$; there are no genuine position eigenstates in quantum field theory.

possibilities is explored further in Chapter 4, and forms the basis of my further development of TI.

In a nutshell, TI treats absorption on the same dynamical footing as emission, providing an unambiguous account of how a "measurement" is finalized, without the infinite regression of apparatus or observers infecting the standard accounts of quantum measurement that neglect absorption. It is also harmonious with relativity (Cramer, 1986, pp. 668–9) and finds support for its even-handed treatment of emission and absorption in quantum field theory, which treats absorption and emission symmetrically.[14] (Emission can be said to be privileged only insofar as it is the starting point for a transaction; something must be created "before" it is destroyed.) Transactions are irreducibly stochastic collapses triggered by absorption events. So in TI, measurements – and any other empirically observable events – are just the results of actualized transactions. There is no need to assign wave functions to macroscopic pointer coordinates, observers, or observer minds; nor, under TI, would this be correct – since an offer wave describes an unabsorbed possibility while macroscopic objects such as pointers and observers are conglomerates of actualized events based on completed transactions.

3.3.2 *Feynman's account of quantum probabilities*

I now examine a presentation by Feynman in his famous Lectures on Physics (Feynman *et al.*, 1964), Vol. 3, in which he explains the rules for calculating probabilities of outcomes by reference to the two-slit experiment (recall Chapter 1). Feynman's presentation, while eminently readable, raises intriguing questions about when or why an experiment is considered "finished," which can be answered in the TI picture. Let's first review his discussion.

Feynman considers a two-slit experiment with electrons, where there is an option to detect which slit each electron went through by shining a light source on the slits. The basic setup, as presented by Feynman, is reproduced in Figure 3.6.

An electron gun emits electrons that can yield interference patterns at the final screen, detected through varying count rates for each position x on the screen. A light source behind the slitted screen emits photons, which can be scattered by the electron into detectors 1 and 2 corresponding to which slit the electron went

[14] In this regard, note that the expression for a quantum field operator associated with a particular spacetime point is a sum of creation (emission) and annihilation (absorption) operators. Cf. Mandl and Shaw, p. 44. Emission of a particle is physically equivalent to absorption of an antiparticle, and *mutatis mutandis*. Absorption is just as important as emission in relativistic theories. It is only in traditional non-relativistic quantum mechanics interpretations that absorption is ignored; TI remedies that discrepancy. This is not to say that TI finds its best relativistic expression in terms of QFT; a more suitable approach is suggested by "direct action" theories such as that of Davies (1970, 1971, 1972). This point is discussed in Chapter 6.

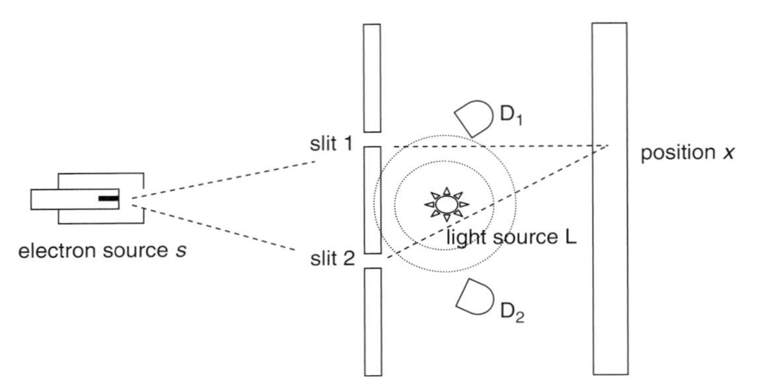

Figure 3.6 The setup for the two-slit experiment with possible "which-slit" detection.

through. The higher the photon's frequency, the smaller its wavelength and the more accurate its which-slit detection. More specifically, the sharp measurement consists of aiming the photon precisely at one of the slits so that it has no chance of intercepting an electron going through the other slit.[15] A fuzzier measurement consists of the photon having some chance of intercepting an electron going through the other slit, even though the photon is not aimed there. Feynman presents a quantitative analysis of this experiment, which I first review in standard terms and then in terms of TI.

The standard account

The amplitude for an electron to go from its source s (the electron gun) to slit 1 is $\langle 1|s\rangle$, and similarly the amplitude for an electron to go from s to slit 2 is $\langle 2|s\rangle$. Feynman highlights the "right to left" character of the notation, in which the emitted state is $|s\rangle$ and the projection of $|s\rangle$ onto $|1\rangle$ has the amplitude $\langle 1|s\rangle$ as discussed in Section 3.2.2. Now, in the absence of any detection at the slits, these are just amplitudes, and each is then multiplied by the amplitude to go from either slit to an arbitrary position x on the screen. Thus, the electron amplitudes to go *from* the source, *by way of* slit 1 or 2, *to* position x are:

[15] This technical detail is glossed over in the Feynman analysis but is not necessary for the point he is making. To take it into account, suppose that the photon is precisely aimed at slit 1, with $a = 1$. If the photon is not detected at D_1, it has been absorbed somewhere else; this is what is represented by D_2 (a more realistic placement would have this "default" detector in the "line of sight" of the photon's aim). Detection of the photon somewhere else indicates that the photon was not scattered by the electron at slit 1, so it definitely went through slit 2. As far as the analysis is concerned, it does not matter at which slit the photon is aimed, so Feynman's original drawing is the most general way to indicate the concepts discussed.

$$\phi_1 = \langle x|1\rangle\langle 1|s\rangle \tag{3.9a}$$

$$\phi_2 = \langle x|2\rangle\langle 2|s\rangle \tag{3.9b}$$

That is, the amplitude to go from the source to slit 1 (or 2) is multiplied by the amplitude to go from slit 1 to the position x on the screen. Feynman notes here that the rule for calculating the probability of an outcome for intermediate unobserved ("indistinguishable") states is to multiply the amplitudes for each step of the process, and then to add those amplitudes for the overall amplitude of the process. Finally, one squares that amplitude to get the probability that an electron starting out from the source ends up at position x, given that both slits are open.

But we're not done yet. The next step in the analysis is to take into account the emission of a photon from light source L each time an electron goes through the apparatus. The photon has a certain amplitude to be scattered into either detector D_1 or D_2. That amplitude depends on the design of the apparatus and the energy of the photons. For example, in an ideal, sharp measurement, the photon will *only* be scattered into D_1 by an electron going through slit 1. But Feynman keeps the analysis general to allow for less precise or "unsharp" measurements. For instance, a photon of low energy has a longer wavelength, which means that it is less localized and therefore gives a less precise measurement of the electron's position than a higher-energy photon.

Suppose we don't specify how sharp the measurement is, and just allow for the possibility that the photon could be scattered into the wrong detector: that is, we allow a non-zero amplitude that the photon could be scattered by an electron going through slit 1 into D_2, and vice versa. Feynman calls the amplitude for scattering the photon into the correct detector (labeled with the same number as the slit) a, and the amplitude for scattering into the wrong detector b. The amplitudes for the total system of electron + photon are the product of the individual amplitudes, so we have (first in words, then in symbols):

Amplitude for photon to go from L to D_1 and electron to go from s to x by either slit = [(amplitude for electron going from s to slit 1) times (photon "correct" amplitude a) times (amplitude for electron going from slit 1 to x)] plus [(amplitude for electron going from s to slit 2) times (photon "incorrect" amplitude b) times (amplitude for electron going from slit 2 to x)] =

$$\langle x|1\rangle a\langle 1|s\rangle + \langle x|2\rangle b\langle 2|s\rangle = a\phi_1 + b\phi_2 \tag{3.10a}$$

and similarly:

Amplitude for photon to go from L to D_2 and electron to go from s to x by either slit =

$$\langle x|2\rangle a\langle 2|s\rangle + \langle x|1\rangle b\langle 1|s\rangle = a\phi_2 + b\phi_1 \tag{3.10b}$$

Now comes the interesting part. The photon can end up in two "distinguishable" states, either at D_1 or D_2 (where I put scare quotes around "distinguishable" since this is what needs ontological disambiguation). If we want the probability that the electron ends up at x and the photon ends up at either detector, then according to the standard account, because the photon detections are "distinguishable," we square the individual amplitudes applying to each photon detector (equations (3.10a) and (3.10b)) and then add those squared quantities:

$$\mathrm{P}(\text{electron at } x,\ \text{photon at } \mathrm{D}_1 \text{ or } \mathrm{D}_2) = |a\phi_1 + b\phi_2|^2 + |a\phi_2 + b\phi_1|^2 \tag{3.11}$$

Now, let us check that we get an interference pattern if the photon "measurement" is maximally fuzzy; that is, if $a = b$ (in this case it works out that their value must be $1/\sqrt{2}$):

$$\mathrm{P}_{\text{fuzzy}}(\text{electron at } x,\ \text{photon at } \mathrm{D}_1 \text{ or } \mathrm{D}_2) = \frac{1}{2}\left(|\phi_1 + \phi_2|^2 + |\phi_2 + \phi_1|^2\right)$$

$$= \frac{1}{2}\left(2|\phi_1|^2 + 2|\phi_2|^2 + 2(\phi_1{}^*\phi_2 + \phi_2{}^*\phi_1)\right)$$

$$= |\phi_1|^2 + |\phi_2|^2 + (\phi_1{}^*\phi_2 + \phi_2{}^*\phi_1) \tag{3.12}$$

which is the same result as if there were no photon at all, and interference is evident in the cross terms.

On the other hand, for a perfectly sharp measurement, $a = 1$ and $b = 0$, so we get

$$\mathrm{P}_{\text{sharp}}(\text{electron at } x,\ \text{photon at } \mathrm{D}_1 \text{or } \mathrm{D}_2) = |\phi_1|^2 + |\phi_2|^2 \tag{3.13}$$

which clearly loses the cross terms and the interference.

I reviewed this discussion by Feynman in the traditional manner because his account of the rules for calculating probabilities raises some interesting questions that I think are well answered in the TI picture. For example, here is what Feynman says about the difference between the conditions requiring (1) adding individual amplitudes before squaring and (2) squaring individual amplitudes first and then adding them:

> Suppose you only want the amplitude that the electron arrives at x, regardless of whether the photon was counted at D_1 or D_2. Should you add the amplitudes [for equations (3.10)]? No! *You must never add amplitudes for different and distinct final states.* Once the photon is accepted by one of the photon counters, we can always determine which alternative occurred if we want, without any further disturbance to the system ... do not add amplitudes for different final conditions, where by "final" we mean at the moment the probability is desired – that is, when the experiment is "finished". You do add the amplitudes for the different *indistinguishable alternatives* inside the experiment, before the complete process is finished. At the end of the process, you may say that "you don't want to look at the photon". That's your business, but you still do not add the amplitudes. Nature does not know what you are looking at, and she behaves the way she is going to behave whether you bother to take down the data or not. (Feynman *et al.*, 1964, Vol. 3, pp. 3–7; original italics and quotations)

But *what* is it that nature is doing that is independent of whether we look or not? What physical circumstance defines when the experiment is "finished"? What makes the two photon states "distinguishable"? What counts as a "disturbance" and what doesn't? These questions are at the very core of the measurement problem and are not answered in the usual pragmatic approaches, which use language like "distinguishable" or "irreversible" without being able to define those conditions in unambiguous physical terms. In particular, according to the usual approach, there is supposedly an ongoing entanglement of the quantum systems with objects in their environment, including measuring apparatus.[16] This is the point of Schrödinger's cat paradox. Despite Feynman's language about nature doing what she does whether or not we are looking, the criteria for when experiments are "finished" inevitably end up referring to the choices of experimenters as to what to measure and/or what can be distinguished *by experimenters*. So, Feynman's obvious (and laudable, in my view) intent to portray the physics as independent of observers and their knowledge sidesteps the fact that the usual account inevitably drags observers back in. (This awkwardness is highlighted by his choice to put "finished" in quotes.) Let's see now how TI resolves this conundrum.

3.3.3 TI as the ontological basis for Feynman's account

First, recall that in TI there is no "measurement" – indeed, no actualized event – *unless confirmation waves (CW) are generated by an absorber.* Let us now revisit equations (3.10). Suppose the photon was aimed at slit 1 – recall note 16 which

[16] The decoherence program is concerned with showing that this alleged entanglement reduces to an approximately classical world, but as noted in Chapter 1, that program depends on assumptions about which part of the universe is the system under study and which is its environment. This makes the account observer-dependent and of course conflicts with Feynman's portrayal of nature as "doing what she does" regardless of our knowledge or decisions. In fact there is no "classical world" in the decoherence approach unless such decisions are made.

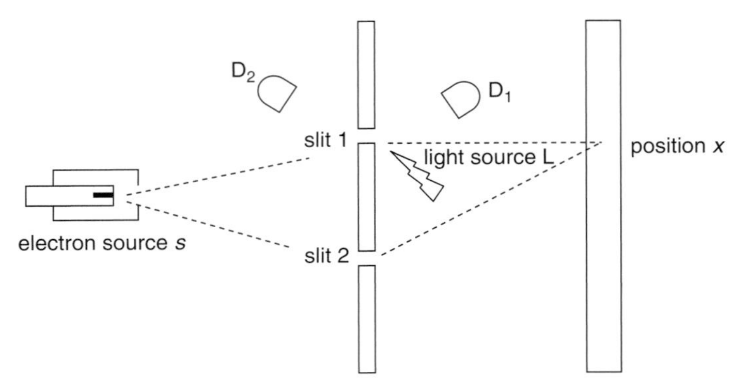

Figure 3.7 The more precise setup showing the light source aimed at slit 1.

discusses the need to take into account that the light source has to be aimed at one or the other slit for a sharp measurement corresponding to detection of the photon at D_1. A good measurement means that the presence of the electron at slit 1 causes the photon to be scattered into D_1. In this case, a detection of the photon at D_2 means that the photon aimed at slit 1 did not encounter the electron and just passed through the slit to detector D_2, which indicates that the electron definitely went by way of slit 2. I illustrate this more precise setup in Figure 3.7.

The interaction between the photon and the electron is illustrated schematically in Figure 3.8. The amplitudes a and b play the part of scattering amplitudes for various incoming and outgoing states of the photon and electron. For a and b arbitrary, and the light source L aimed at slit 1, we have the possible scattering events shown in Figure 3.8 corresponding to equation (3.10a).

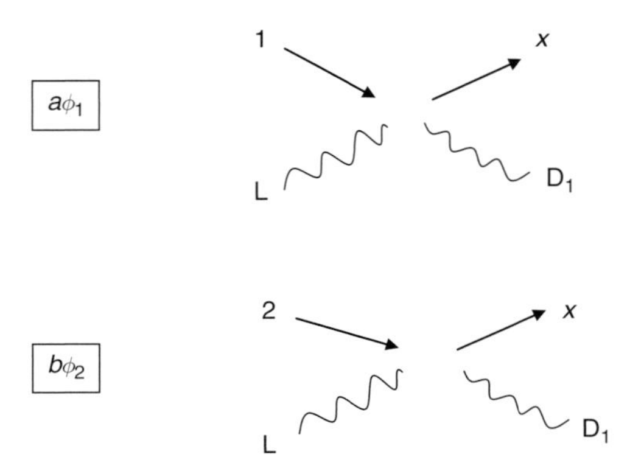

Figure 3.8 The two scattering amplitudes contributing to activation of detector D_1 for the photon for an imperfect measurement. The top diagram corresponds to the correct photon detection and the bottom diagram corresponds to the incorrect photon detection. The electron is represented by the straight line and the photon by the wave line.

In calculating the amplitude for a scattering process, one considers all the different ways (up to a given order) that a particular set of events can occur. In this case, we are interested in all the ways that an electron and photon can start out in states $|s\rangle$ and $|L\rangle$, respectively, and end up in states $|x\rangle$ and $|D_1\rangle$, respectively. I suppress the electron's initial state $|s\rangle$ and just indicate whether it has "gone through" slit 1 or 2 by the corresponding number. The electron amplitude for the top diagram is ϕ_1, and for the bottom diagram it is ϕ_2. The photon amplitude for the top diagram is a, and for the bottom diagram is b. The two-particle amplitude for each diagram is the product of the individual particles' amplitudes, and the total amplitude for the scattering process is the sum of the two amplitudes for the diagrams.

Now let's consider the foregoing in the TI picture. The offer wave corresponding to "electron goes through either slit, photon goes to D_1" is a superposition of the two components illustrated in Figure 3.8. The only "distinguishable" events are the photon detection at D_1 and the electron detection at x, and this is because a composite (electron and photon) confirmation wave was generated corresponding to the electron's offer wave interacting with detector x and the photon's offer wave interacting with detector D_1. The confirmation wave's amplitude is the complex conjugate of the superposition of both diagrams, i.e.

$$\text{Amp}[\text{CW}(x, D_1)] = a^*\phi_1^* + b^*\phi_2^*$$

Recalling Section 3.2.3, the probability of the event in question is just the product of the amplitudes of the OW and CW for the event, or the absolute square of the OW amplitude. (Note that we will have interference in this case, due to the fact that b is different from zero.) The reason that we square the sum of the amplitudes for the D_1 photon detection is precisely because *there is a CW from detector D_1* (as well as from the electron detector at x). In contrast, there was never a CW from either slit, so those electron states are indistinguishable. Indistinguishability means, in TI, that no CW is generated for a particular OW component.

Now, let us see what happens when $a = 1$ and $b = 0$. In this case, the amplitude for electron detection at x and photon detection at D_1 is given only by the top diagram in Figure 3.8.

There are CW generated at x and at D_1, so we square the amplitude for the diagram:

$$\text{P}(a = 1, \text{electron at } x, \text{ photon at } D_1) = |\phi_1|^2 = \langle x|1\rangle\langle 1|s\rangle\langle s|1\rangle\langle 1|x\rangle$$

which is the probability distribution for an electron going through slit 1 (as if slit 2 were closed). Interference is lost in this case, even though there are no CWs associated with the slit. This is because the only available transactions (i.e., OW with CW responses) are those projecting the total system onto *either* D_1 and x by way of slit 1 *or* D_2 and x by way of slit 2.

The physical content behind the above probabilities is that the squared quantities represent confirmations of the corresponding offer wave components (recall Section 3.2.3). Now recall that the standard account lacks a basis for considering certain events "distinguishable" and others "indistinguishable." The TI picture simply identifies distinguishable events as those corresponding to *confirmed* offer waves in the form of squared amplitudes; indistinguishable events are those represented only by unconfirmed offer wave amplitudes, as for example the electron OW component $\langle 1|s\rangle$ when a is less than 1. But it is important to keep in mind that the confirmation of an OW does not necessarily mean that the event or property corresponding to that particular OW is actualized; rather, it means that *there is a determinate fact of the matter* as to whether that property or another property corresponding to the same observable is actualized. In other terms, it means that the usual classical rules of probability apply to the situation – i.e., the "cat is either alive or dead," in contrast to the fuzzy quantum logic of saying "the cat is in a superposition of alive and dead."

Note that this account obviates the need to refer to observer-dependent criteria such as (in Feynman's phrasing) "the moment the probability is desired." It allows us to take away the scare quotes from Feynman's reference to the experiment being "finished"; the experiment is unambiguously finished when a confirmation is generated that allows us to apply the rules of classical probability; i.e., to say that there is a definite fact of the matter about the whereabouts of both particles in the experiment. Those rules apply, not because someone desired a probability, but because there was a *physical process* in play: a confirmation, which brought about a determinate event. The confirmation is what creates the "disturbance" that disrupts the quantum superposition. TI allows us to define what "disturbance" really means, in concrete physical terms.

One last detail needs to be addressed: the reader may worry about the apparent asymmetry of having to choose a slit at which to aim the photon, which in this discussion was arbitrarily chosen as slit 1. To see that this is not a problem (i.e., no generality is lost), let's return to equation (3.10b) for the amplitude of an electron at x and a photon at D_2:

$$\langle x|2\rangle a\langle 2|s\rangle + \langle x|1\rangle b\langle 1|s\rangle = a\phi_2 + b\phi_1 \tag{3.10b}$$

Note that this holds regardless of which slit is targeted by the photon source. If D_2 is the "null" detector – i.e., a photon is detected there only if it failed to interact with an election – a is still the amplitude that an electron going through slit 2 will correctly result in the photon being detected at D_2. This is because an electron going through slit 2 has very little chance of scattering the photon aimed at slit 1 (where that chance depends on the magnitude of the "error" amplitude b) and so it will just continue on to D_2 "by default." The photon won't be scattered into D_2 by the electron, but what matters for the calculation is not *how* the photon gets to its destination but rather the amplitude that it will do so, and the correspondence of that amplitude with the state of the electron. Even if the photon is aimed at slit 1, it correctly gets to D_2 for an electron state corresponding to slit 2. On the other hand, if the photon ends up at D_1 for an electron state corresponding to slit 2, it's only because the measurement is a "fuzzy" one in which the photon is poorly localized and therefore has a chance of being scattered into D_1 even by an electron going through the "wrong" slit.

3.4 Circumstances of CW generation

As will be explicated in Chapter 6, it is proposed herein that the coupling amplitudes between interacting fields in the relativistic domain are to be identified as the amplitudes for the generation of confirmation waves. This means that the generation of CW is a stochastic process. Thus, although TI can provide a definitive account of "disturbance," that does not translate into a causal, predictable account of disturbance. At the relativistic level, there is no way to predict with certainty whether, for any given instance, a CW will be generated. Couplings between (relativistic) quantum fields give only amplitudes for confirmations.[17] The non-relativistic regime is defined by assuming that all absorbers definitely generate confirmations, since that domain does not involve field couplings.

We can only know after the fact that a CW has been generated. However, in macroscopic situations, detectors are composed of huge numbers of individual potential absorbers, and this assures the generation of CW. In fact, the identification of coupling amplitudes as amplitudes for confirmations allows us to specify an unambiguous physical basis for the notorious "micro/macro" boundary. The macroscopic world is simply the level at which CW are virtually assured and quantum superpositions are thereby collapsed (through actualized transactions).

For example, suppose we want to detect photons by way of the photoelectric effect (in which electrons are ejected from the surface of a metal by absorbing the energy of an incoming photon). In this case, the electron serves as the potential absorber for the photon's OW. The coupling of any individual photon and electron has an amplitude of ~ 1/137, about 0.007, so the probability that an interaction

[17] This fact does not conflict with the TI account of the Feynman experiment, which applies to photon detections at D_1 or D_2 *in coincidence* with electron detections at x. For whenever a photon is detected at D_1 or D_2 it is because a confirmation was in fact generated.

between the two will generate a CW is ~ $(0.007)^2 = 0.00005$, a very small number. Indeed, the probability that a CW will *not* be generated in an interaction of a photon with a single electron is $1 - 0.00005 = 0.99995$, so for any individual electron we may confidently predict that a CW will not be generated. However, consider a small macroscopic sample of metal, say 1 cubic centimeter. A typical metal has roughly 10^{23} free (conduction) electrons in this size sample. The probability that a CW will not be generated in the sample is the product of the probabilities of *every* electron in the sample not generating a CW, or (0.99995) raised to the 10^{23} power, which is an infinitesimal number (your calculator will give zero). Thus it is virtually certain that in any interaction of a photon with a macroscopic quantity of electrons, at least one of them will generate a CW. So when we deal with a macroscopic detector, it is virtually certain that an OW will generate a "response of the absorber" somewhere in the detector, thus physically warranting the Born Rule's squaring procedure.

The preceding explains why we have to work so hard to retain quantum superpositions. In order to obtain observable phenomena we must work with macroscopic quantities of matter, but this is the level at which confirmations (absorber responses) become virtually certain, and the latter are what cause the "collapse" of superpositions of quantum states. With increasing technological sophistication comes the ability to create superpositions of mesoscopic objects such as "buckeyballs" (cf. Arndt and Zeilinger, 2003).

4

The new TI: possibilist transactional interpretation

4.1 Why PTI?

The 1986 version of TI faced some difficulties: (1) the interpretation of multi-particle offer and confirmation waves; (2) the nature of the process leading to an actualized transaction; (3) the apparent possibility of causal loops leading either to inconsistency or inconclusive quantitative predictions.[1] The issue to which (1) refers is the following: in mathematical terms, multi-particle states for N particles are actually elements of a $3N$-dimensional space.[2] For example, a general quantum state for two identical particles A and B contains components A_x, A_y, A_z, B_x, B_y, B_z. So a realist interpretation of such states cannot claim that their offer and confirmation waves exist entirely in ordinary space, but instead must allow for a real "higher", extra-empirical space.[3] The updated version of TI presented herein, possibilist TI or PTI, addresses this by proposing that offer and confirmation waves represent dynamically efficacious *possibilities* whose collective structure constitutes just such a "higher space," which (anticipating the relativistic domain) I call "pre-spacetime" (PST).[4] This is a form of structural realism, since I do not claim to know the material nature (if any) of these possibilities, but rather claim that the formal structure of the theory reflects an existing structure in the real world, albeit an extra-empirical one.

Concerning (2), the earlier version of TI referred to an "echoing" process which, given the higher-dimensional entities involved, should not be thought of as a

[1] I refer here to Maudlin (2002, pp. 199–200) and Berkovitz (2002, 2008).

[2] I use the term "particle" here as a convenience because that is the usual language used in this context. However, the "particles" involved should not be thought of as localized, classical particles. They are just quanta of one or more fields.

[3] In referring to certain entities as "living" in a space, I do not mean to imply that such a space necessarily exists independently as a substance. That is, this locution is not meant to endorse substantivalism about such a space. It just refers to the mathematical characteristics of the manifold of the entities in question.

[4] At the relativistic level, where particles are being created and destroyed and therefore particle numbers are changing, the relevant "higher space" is described by Fock space, the relativistic extension of Hilbert space.

process taking place within the spacetime manifold. Since there is no causal (in the sense of deterministic) way to account for the actualization of one transaction out of several incipient ones, such an actualization is irreducibly stochastic in a way that is not compatible with any causal process within the confines of ordinary spacetime. PTI takes the actualization of a particular transaction (from among a collection of *N* incipient ones) as an extra-spatiotemporal process, more akin to spontaneous symmetry breaking (SSB) than to a back-and-forth dynamical process within spacetime such as the "echoing" of Cramer's original TI. This suggestion will be discussed further below. With regard to (3), issues involving causal loops and possible deviations from the predictions of quantum mechanics can arise only if all the processes involved are thought of as taking place in a "block world," which is not the case in PTI. Thus PTI is not subject to these difficulties, as will be explicated in later chapters.

4.2 Basic concepts of PTI

First, here are the defining characteristics of PTI.

4.2.1 Offer and confirmation waves are physically real, but sub-empirical, possibilities

OW and CW (see Chapter 3) are interpreted ontologically in PTI as *physically real possibilities*. In this context, "real" means physically efficacious but not necessarily *actualized*. (This distinction is elaborated in detail in Chapter 7.) Again, think of the submerged portion of the iceberg in Figure 1.1: from the vantage point of the deck of a ship (representing the empirical realm), we cannot see the submerged portion, but it certainly supports the visible portion and therefore cannot be dismissed as "abstract" or "unreal." In Bohr's words (recall Chapter 2), these entities "transcend the spacetime construct"; however, rather than dismiss them as Bohr did, I allow that they are physically real, even if sub-empirical.

OW and CW are necessary but not sufficient conditions for an actualized event. The remaining necessary condition for an actualized event is that one particular transaction be actualized from a set of *N* incipient ones, where *N* labels the number of absorbers returning CW to the emitter. The adjective "real" thus designates a weaker ontological status than "actual" (i.e., "actual" events are special subsets of real events).[5] An event that is actual is also real. But a real process is not necessarily an actualized process; it may be a possible process.

[5] This is roughly analogous to Lewis' treatment, except for the fact that he considers the "actual" designation as merely indexical. In PTI, an actual event has a different ontological status from a possible (real) event in that

4.2.2 Emission and absorption of quanta occur in an extra-empirical, pre-spatiotemporal realm

Emissions of offer waves and absorptions generating confirmation waves are primary dynamical events which take place in the pre-spatiotemporal realm (PST). Metaphorically speaking, PST corresponds to the submerged portion of the iceberg, and the ocean in Figure 1.1 (the ocean can be thought of as the vacuum state). More precisely, PST is the manifold mathematically described by Hilbert space,[6] the domain of sub-empirical (meaning not directly observable) quantum theoretical objects such as the entities described by quantum states.

If the idea of viewing the realm described by Hilbert space as real (what philosophers term "reifying" Hilbert space) seems strange, it needs to be kept in mind that many of the quantum objects described mathematically by Hilbert space quantities and routinely assumed as existing *somewhere*, cannot be thought of as existing within a spacetime manifold – in the sense of being localized at a point or within any spacetime region. For example, the ubiquitous "vacuum state" or ground state in the energy representation, $|0\rangle$, has no spacetime arguments (i.e., it is not a function of x, y, z, t) and cannot be considered to exist in any well-defined region within spacetime. (For a technical account of why this is so, a good place to start is Redhead's "More ado about nothing" (1995).) So, in a pragmatic sense, physicists already take objects described by Hilbert space quantities as real. When pressed, they might respond that "well, of course it's not *real*, because it does not exist in spacetime"; but this merely expresses the conventional, often unexamined definition of "real" that insists: "real = existing in spacetime," which is precisely what I claim needs to be questioned. The point is that these objects are assumed to be physically efficacious: they are acted upon by other physically efficacious objects (such as operators, which likewise do not exist in spacetime) and they can give rise to concrete observations (e.g., measurement outcomes). Thus physicists already view them as essentially real, even though they do not exist in spacetime.[7]

It is important to note that emitters are generally only offer waves themselves – i.e., interacting fields which couple to other fields. A macroscopic emitter is a collection of linked actualized transactions, as is any observable macroscopic object (this notion of linked transactions will be further clarified in Chapters 7 and 8).

the former exists in spacetime (the "tip of the iceberg") whereas the latter exists in PST (the submerged portion of the iceberg). In this respect, the "many worlds" of Everettian interpretations can be considered to correspond to Lewisian "possible worlds." This issue is addressed further in Chapter 7.

[6] Technically, this is really Fock space, the relativistic extension of Hilbert space (see note 4).

[7] Another approach consists in taking quantum objects as epistemic (i.e., referring primarily to the knowledge and/or intentions – e.g., of what physical quantity to measure – of an observer). This approach has been critiqued earlier (Chapter 1, note 20 and in Chapter 2).

A macroscopic object termed an "emitter" is just a type of actualized object[8] with a high probability of generating fields which can couple to other fields (i.e., microscopic emitters and absorbers). An example is an electrode exposed to an electromagnetic field strong enough to eject its electrons through the photoelectric effect. The ejected electrons are offer waves, but the atoms that emitted them are, in general, "just" offer waves as well.[9]

If the previous sentence seems surprising, remember that Ernst Mach railed against the idea that atoms were "real" because they were not directly observable. We have become so accustomed to the concept of atoms that we have forgotten that they are not directly observable: we never really "see" an atom. We can image small numbers of atoms through interactions with a scanning tunneling electron microscope (STM), but those images result from transactions between electron offer waves emitted by the STM and absorbing atoms in the imaged sample. The transactions result in a measurable current, and there are variations in this current (fewer or more transactions) depending on how many or few atoms comprise a given portion of the sample. The changes in current are rastered onto a 2-dimensional surface to yield a kind of "image" of the scanned surface, with regions corresponding to larger currents being identified with atoms. Thus we are not really "seeing atoms," we are seeing a representation of changes in the transacted current due to interactions between an electron offer wave current and atomic absorbers.

Returning to our basic macroscopic emitter, the apparent solidity of the electrode composed of the atoms is based upon transactions among the atoms (through interatomic forces) and between the atoms and other absorbers (for example, those in our hands or eyes). This transactional basis of sensory perception is illustrated in Figure 4.1, which shows a man viewing and touching a table (offer waves indicated in black and confirmation waves in gray; remember, though, that these are not entities propagating in spacetime; they are extra-empirical). I will return in Chapter 7 to the epistemic (knowledge-based) implications of this account with reference to the enigmatic and elusive "table-in-itself" discussed at length in Bertrand Russell's *The Problems of Philosophy* (1959, chapter 1).

[8] Fields (e.g., 2011) is skeptical of the idea that such objects can be considered "emergent" or well-defined absent a classical distinction between objects, a distinction which requires an observer. However, for present purposes, I don't need to specify where the "macroscopic emitter" ends and (say) the air surrounding it begins. When humans manipulate macroscopic emitters, they are interacting with a physical entity with a high probability of emitting offer waves, and that is all that is required for this account. That is, the physical entities comprising the "air" portion have a drastically lower probability of emitting offer waves than the physical entities comprising the "laser" portion.

[9] Technically, the source of a field is a current, which has units of energy and is proportional to the square of the usual quantum state. This is a feature of relativistic quantum mechanics and takes into account that when a field quantum is emitted, the original incoming "emitter" state is modified and becomes an outgoing state.

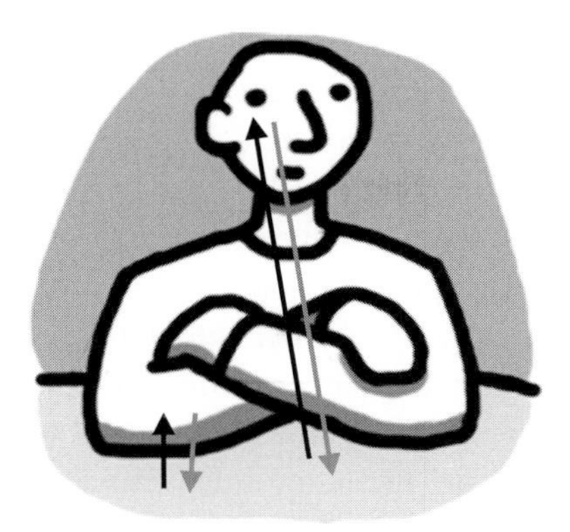

Figure 4.1 Macroscopic objects are perceived via transactions between offer waves emitted by components of the object and confirmations generated by absorbers in our sense organs.

4.2.3 Incipient transactions are established through OW–CW encounters in PST

In a generalization of the Wheeler–Feynman approach discussed in Chapter 3 (upon which the original TI was based), PTI (as well as TI) proposes that an absorber, in coupling with an emitted offer wave, generates a confirmation wave.[10] This process can be viewed as a generalization of Newton's first law, which observes that, in the classical domain, a mass acted upon by a force **F** exerts an equal and opposite force –**F**. In general, there will be more than one absorber A_i ($i = \{1, N\}$) for an emitted offer wave $|\Psi\rangle$, and in such cases the latter is then projected into components corresponding to the capabilities of each absorber. Formally,

$$|\Psi\rangle = \sum_{i=1}^{N} |A_i\rangle\langle A_i|\Psi\rangle = \sum_{i=1}^{N} \langle A_i|\Psi\rangle|A_i\rangle \tag{4.1}$$

which reflects the usual projection of a given state $|\Psi\rangle$ onto a particular basis. Thus, PTI provides a physical referent for a common mathematical expression (4.1) in the theory. This interpretation removes the arbitrariness of basis often associated with quantum states. That apparent arbitrariness arises because a crucial aspect of the mathematical formalism has remained physically unapplied in standard

[10] Taking into account the relativistic domain, one should say that an absorber has an *amplitude* to emit a confirmation wave (just as an emitter has an amplitude to emit an offer wave). This is explored further in Chapter 6.

accounts of quantum mechanics. Here, we see that the appropriate basis is *physically* determined by the availability of a set of absorbers.[11]

The arbitrariness of basis due to neglecting absorption is analogous to the underdetermination of the force that will be experienced by an object O moving at speed s toward another object O′ when the speed of O′ has not been specified (thus leaving their relative speed unspecified). Just as in classical situations involving Newton's first law, dynamical interactions take place in encounters between OW and CW. A falling object encountering a table will feel a responding force and undergo compression; similarly, an offer wave meeting a confirmation wave will precipitate an incipient transaction which may be actualized. In terms of (4.1), such encounters are represented by the weighted projection operators $|\langle A_i|\Psi\rangle|^2|A_i\rangle\langle A_i|$. This expression is the product of two factors: the matching (projected or attenuated) component of the original offer wave, $\langle A_i|\Psi\rangle|A_i\rangle$, and the resulting confirmation wave

$$\langle A_i|\Psi\rangle^*\langle A_i| = \langle\Psi|A_i\rangle\langle A_i|$$

which is received by the emitter and projected onto its state $|\Psi\rangle$ in the advanced ("brac") version of (4.1), represented in (4.2) below.[12] Here I use the fact that $1 = \sum_j |\Psi_j\rangle\langle\Psi_j|$, where the emitter state $|\Psi\rangle = |\Psi_J\rangle$ corresponds to a particular value J of the index j:

$$\langle A_i| = \sum_{j=1}^{M} \langle A_i|\Psi_j\rangle\langle\Psi_j| \qquad (4.2)$$

In the original TI, this type of process was referred to as taking place in "pseudotime," where the latter was a heuristic device. In PTI, this process is fully extra-spatiotemporal; it takes place in the realm of dynamical, pre-spacetime possibilities described by Hilbert space. When an absorber generates a CW, it also emits a matching OW (the ket $|A_i\rangle$), just as in the original TI. Likewise, the emitter also emits an advanced component, $\langle\Psi_J|$. Recall in this regard that the original presentation of TI in Cramer (1986) made use of the Wheeler–Feynman time-symmetric formulation of electrodynamics as an analogical springboard for an interpretation applying to the quantum

[11] It is theoretically possible to have an incomplete set of absorbers (e.g., an absorber for a "spin-up" state along some direction z but no absorber for "spin-down"). This issue (which does not negate the fact that a basis is determined with respect to absorbers) is discussed in Chapter 5.

[12] To get the squared amplitude expression (Born Rule), one can either simply multiply the OW and CW amplitudes together as presented above, or follow the entire process of the original OW as it is attenuated, "reflected" back as a complex-conjugated CW, and is finally projected onto the original emitter state, resulting in a "round-trip" amplitude reflecting the Born Rule.

domain, by arguing that such a "remnant" retarded (future-directed) wave from an absorber would be canceled due to its being out of phase with the original OW component. The original TI can be viewed as the semi-classical, single-particle limit of PTI, in which one can disregard the proliferation of Hilbert space dimensions required for dealing with systems of more than one particle. The possibility of giving an account of the "cancellation" of residual advanced effects from the emitter and retarded effects from the absorber, as provided in Cramer (1986), demonstrates the capability of the interpretation to accommodate the correspondence principle, which requires that the quantum account be observationally consistent with classical predictions.

4.2.4 Spacetime is the set of actualized transactions

In this interpretation, spacetime is no more – and no less – than the set of actualized transactions. Thus, actualizations of transactions based on OW and CW superpositions give rise to the set of related events comprising the spacetime theater. In an actualized transaction, the emission defines the past and the absorption defines the future. That is, *past and future supervene on actualized transactions; there is no "spacetime" without actualized transactions*. The apparent 4-dimensional spacetime universe is not something "already there"; rather, it crystallizes from an indeterminate (but real) PST of dynamical possibility. Thus, spacetime "grows" but not in the usual "growing universe" sense wherein an advancing "now" proceeds from present to future; rather, events arise from a set of dimensions (the Hilbert or Fock space manifold) outside spacetime. In fact, it is the past that "grows" and is extruded from the present; in PTI there is no actualized future. This picture is explored further in Chapter 8.

If, as a contingent fact of our universe, all emitted offer waves are provided with at least one absorber (PST is "opaque"), then all emitters must ultimately exist within spacetime since they will participate in an actualized transaction. However, absorbers corresponding to unactualized transactions – that is, transactions that "failed" – do not exist within spacetime (unless the same absorber participates in an actualized transaction from some other offer wave/emission).[13]

Again, it needs to be kept in mind that the "untransacted" absorbers are considered in PTI to be real possibilities, so that the terminology "does not exist within spacetime" means unmanifest within the observable "spacetime theater,"

[13] Thus the apparent temporal asymmetry we observe in the universe may be attributable to the inevitable fact that creation (that is, emission of a positive energy state) necessarily precedes annihilation; one cannot annihilate unless there is *something already there* to annihilate. This fact is reflected in quantum field theory in the asymmetry in the action of creation and annihilation operators on the vacuum state $|0\rangle$ (which designates that no quanta are present). If you try to operate on the vacuum state with the annihilation operator $\mathbf{a}$, you end up with literally nothing; not even a vacuum state! That is, $\mathbf{a}|0\rangle = 0$. In contrast, you can act on any state, including the vacuum state, with the creation operator $\mathbf{a}^\dagger$ and still have a well-defined state.

Figure 4.2 A geode is a roughly spherical pocket of crystals growing in a shell of amorphous material. It is built up through mineral deposits in water flowing through lava bubbles and other hollow structures.
Source: http://en.wikipedia.org/wiki/File:Geode_angle_300×267.jpg

Figure 4.3 The set of events in spacetime emerges from a pre-spacetime realm of indeterminate possibility, as the inner ordered, crystalline structure of a geode arises within an outer shell of amorphous mineral. Pictured is Javier Garcia Guinea inside the huge geode he discovered in Almeria, Spain.
Source: Private collection of J. Garcia Guinea, 2002; used with permission

yet still physically real; PTI is realist about possibilities. Thus spacetime arises from beyond itself, from roots of possibility in PST. If this picture seems strange or hard to visualize, it can be considered roughly analogous to the formation of a geode (see Figures 4.2 and 4.3). Strictly speaking, a geode forms through the depositing of minerals from surrounding water into a hollow bubble of lava. But if the source of the mineral deposits were in the lava "shell," then the analogy would be closer.

Figure 4.4 The circular structure does not exist without the people comprising it. [Pictured is a painting by Hans Thoma, *Der Kinderreigen, 1884.*]

If we think of the geode formation in this latter way, it is an outer "shell" of possibilities which surrounds and gives rise to the crystallized events of the spacetime theater.

More precisely, if the crystals are gradually built up from just inside the shell, that inner layer of shell represents the present, or "now," as experienced by an observer whose sense organs are absorbers on the "receiving" end of a transaction (as in Figure 4.1). The crystalline structure growing toward the center of the geode interior represents the actualized past that continually grows from the now. The outer amorphous shell represents physically real but sub-empirical content outside this spatial realm in a "higher space" of possibilities.[14] I discuss this metaphysical picture in more detail in Chapters 7 and 8.

At this point, I should touch base with the philosophical terminology for the view of spacetime presented in the preceding section: it is known as *relationalism* or *antisubstantivalism*. This is the view that spacetime does not exist as a substance or as a background "container" for events. Instead, the term "spacetime" describes the structured set of events themselves. This view can be illustrated by reference to Figure 4.4, which shows a group of people forming a circle.

[14] Technically, the shell represents entities in Hilbert space (or Fock space in the relativistic domain, for non-interacting fields).

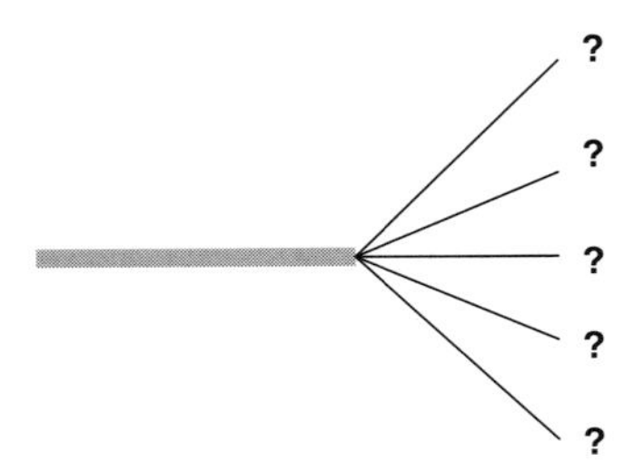

Figure 4.5 Spontaneous symmetry breaking: a transformation of a theory component in which a multiplicity of states or outcomes is possible, none of which can be "picked out" by anything in the theory as the realized state or outcome.

A circular structure exists, but it only exists by virtue of the people comprising it. In the same way, according to relationalism, spacetime only exists by virtue of the events comprising it.

4.3 Addressing some concerns

Let us now return to concerns (2) and (3) in a little more detail. (Concern (1) is immediately resolved in PTI by positing that quantum state vectors or wave functions represent multi-dimensional possibilities whose realm is PST, not ordinary spacetime.)

4.3.1 How a transaction forms

Recall that the subject of concern (2) is that the "pseudotime" process of the original TI does not seem to fully account for why or how a particular transaction is actualized while others are not. If we take the domain of transaction formation as PST rather than spacetime, then an account cannot be given in terms of any causal process *within* spacetime in the usual sense – i.e., along or within light cones, since the latter are confined to spacetime. Instead we need to turn to a similar situation in physics in which there are apparently many possibilities but only one is realized: "spontaneous symmetry breaking" (SSB).

In SSB, the governing theory for the phenomena under study specifies a symmetric situation, illustrated schematically in Figure 4.5. A component of the theory (e.g., a field) undergoes a transformation in which a multiplicity of states or outcomes is possible, none of which can be "picked out" by anything in the theory as the realized state or outcome.

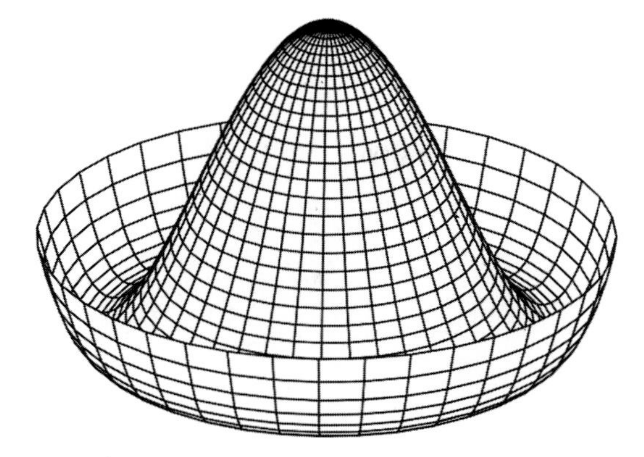

Figure 4.6 The "Mexican Hat" potential which creates an infinite number of possible ground states in the Higgs *et al.* mechanism.
Source: http://en.wikipedia.org/wiki/File:Hans_Thoma_003.jpg

A specific example of this phenomenon occurs in the "Higgs mechanism,"[15] in what is termed the "Standard Model" of elementary particle theory. According to this widely accepted model pioneered by Steven Weinberg and Abdus Salam, the quanta of some force-carrying fields acquire a mass by way of a process in which the ground (vacuum) state of the field undergoes the kind of transformation conceptually depicted above. What was a single vacuum state of the field acquires what is termed a "degeneracy" – that is, many possible ground states (in fact, an infinite number of them). This situation is illustrated in Figure 4.6; the symmetry breaking occurs through what is called a "Mexican hat" potential due to its shape. The original ground state becomes unstable and corresponds to the crown of the "hat"; the infinite set of ground states is found all around the ring at the lowest point. The theory does not provide any way of deciding which of these many ground states is realized. But, according to the theory, the fact that the quanta in question have a non-zero mass indicates that one has been realized.

4.3.2 Curie's principle and Curie's extended principle

The situation just described seems to run afoul of a philosophical doctrine[16] termed "Curie's principle" in honor of Pierre Curie, who championed it. (The principle is actually a version of Leibniz' "principle of sufficient reason" (PSR), which states

[15] The idea was actually arrived at independently in 1964 by Peter Higgs; Robert Brout and Francois Englert; and Gerald Guralnik, C. R. Hagen and Tom Kibble.

[16] Referring to something as a "philosophical doctrine" simply means that it is presumed to be true on the basis of certain metaphysical or epistemological beliefs or principles. Modern physical theory could be taken as indicating that the PSR may not be applicable to the physical world, however compelling it may seem to those who have championed it.

Figure 4.7 A political cartoon (ca. 1900) satirizing US Congress's inability to choose between a canal through Panama or Nicaragua, by reference to Buridan's ass. *Source*: W. A. Rogers, New York Herald (Credit: The Granger Collection, NY).

that any event occurs for a reason or cause which specifies or determines that event, as opposed to some other event. The PSR implies that, absent such a reason or cause, the event in question will not occur.)

Curie's principle states that an asymmetric result (i.e., the choice of one outcome among many equally possible ones) requires an asymmetric cause. That is, it holds that there can be no sound basis for saying that one of the outcomes "just happens"; one must be able to point to a definite reason for that outcome (the reason being the asymmetric cause). This principle is illustrated by a humorous paradox, "Buridan's ass," discussed by French philosopher Jean Buridan, in which a hungry donkey is placed between two equally distant, identical bundles of hay (see Figure 4.7). According to an implicit version of Curie's principle being satirized by Buridan,[17] the donkey will starve to death because it has *no reason* to choose one pile of hay over the other. Of course, our common sense tells us that the donkey will find a way to begin eating hay, even though one can provide no reason for it (hence the paradox). Similarly, in SSB, the field in question arrives in a particular ground state though no specific cause for that choice can be identified. If we take

[17] Buridan was satirizing the doctrine of moral determinism, which views a person's moral actions and choices as fully determined by past events.

Curie's principle to be applicable to the above, then it appears that nature simply violates the principle (as does a hungry donkey).[18]

There is another way of looking at this situation, described by Stewart and Golubitsky (1992). These authors point out that nature seems to be replete with symmetries that are spontaneously "broken," similar to the way in which the symmetry of the vacuum state is broken by the Higgs *et al.* mechanism. In general, a symmetrical system may, under certain circumstances, be capable of occupying any one of a set of symmetrically related states, with no particular state being privileged; thus the particular state in which it happens to be found is arbitrary. Stewart and Golubitsky therefore suggest that nature conforms to a weakened version of Curie's principle, which they call the "extended Curie's principle": "physically realizable states of a symmetric system come in bunches, related to each other by symmetry"; or, alternatively, "a symmetric cause produces one from a symmetrically related *set* of effects" (original italics; Stewart and Golubitsky, 1992, p. 60). Technically, the "bunches" are subgroups of the original symmetry group which has been "broken" by the dynamical situation under consideration.

As noted by Stewart and Golubitsky, a famous illustration of symmetry breaking appears in the iconic 1957 photograph of the splash of a milk droplet by high-speed photography pioneer Harold Edgerton. I reproduce it here in Figure 4.8. The authors point out that the pool of milk and the droplet both have circular symmetry, but the "crown" shape of the splash does not – it has the lesser symmetry of a 24-sided polygon. This happens because the ring of milk that rises in the splash reaches an unstable point – a point where the sheet of liquid cannot become any thinner – and "buckles" into discrete clumps (the laws of fluid dynamics must be used to predict that there are 24 clumps). But the locations of the clumps are arbitrary; that is, the clump appearing just beneath the white droplet above the crown could just as well have been a few degrees to the left (with all the other clumps being shifted by the same amount). There is thus an infinite number of such crowns possible, but only one of them is realized in any particular splash.

Thus the authors point out that, while the mathematics describing a particular situation may provide for a large, even infinite, number of possible states for a system to occupy, in the actual world only one of these states can be realized. They put it this way:

> A buckling sphere can't buckle into two shapes at the same time. So, while the full potentiality of possible states retains complete symmetry, what we observe seems to break

[18] Is there a volitional basis for actualization? Buridan's ass is hungry, so he chooses to eat one of the piles of hay, even if there is no "reason" for it. Does nature then express a certain volitional capacity? Or, put another way, could such an uncaused "choice" be seen as evidence of the creativity of nature?

Figure 4.8 Milk drop coronet.
Source: Harold E. Edgerton, Milk-Drop Coronet, 1957. © 2010 Massachusetts Institute of Technology. Courtesy of MIT Museum

it. A coin has two symmetrically related sides, but when you toss it it has to end up either heads or tails: not both. Flipping the coin breaks its flip symmetry: *the actual breaks the symmetry of the potential.*

(Stewart and Golubitsky, 1992, p. 60)

I have italicized the last sentence because it expresses the same deep principle underlying the PTI picture: mathematical descriptions of nature, with their high degree of symmetry, in general describe a *set of possibilities* rather than a specific state of affairs. Nevertheless, the astute reader may well raise the following question: but isn't it the case that, in the classical domain, we can always find some external influence, however small, that *caused* the system to end up in one particular state as opposed to some other possible state? This would seem to apply, for example, in classical chaotic systems such as the double pendulum (see Figure 4.9). For large initial momentum, such a system's set of possible trajectories encounter "bifurcation points" (essentially, "forks in the road") in

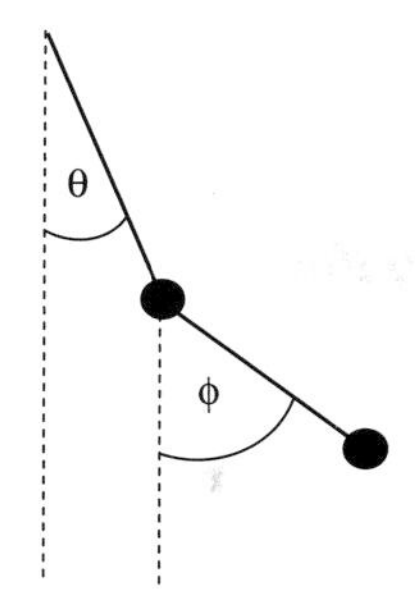

Figure 4.9 A double pendulum, whose classically described motion encounters bifurcation points.

which a specific choice of trajectory is sensitive to perturbations down to the Planck scale (i.e., random quantum fluctuations).

The authors address this, at least in part, as follows:

> we said that *mathematically* the laws that apply to symmetric systems can sometimes predict not just a single effect, but a whole set of symmetrically related effects. However, Mother Nature has to *choose* which of those effects she wants to implement.
>
> How does she choose?
>
> The answer seems to be: imperfections. Nature is never perfectly symmetric. Nature's circles always have tiny dents and bumps. There are always tiny fluctuations, such as the thermal vibration of molecules. These tiny imperfections load nature's dice in favor of one or the other of the set of possible effects that the mathematics of perfect symmetry considers to be equally possible.
>
> *(Stewart and Golubitsky, 1992, p. 15)*

Thus, the apparent answer of the authors to the question of what causes the system to end up in a particular state is: quantum fluctuations. That is, the cause is found *outside* the mathematical formulation of the set of possible solutions for the classical system (or, in the case of certain chaotic systems such as the double pendulum, by following the classical account into the quantum domain in which its deterministic aspect breaks down). It appears that, strictly speaking, when considering symmetry breaking in the classical domain, one could always point to some *external* cause of this type, even if only a random quantum fluctuation. So when the authors say that "the actual breaks the symmetry of the potential," they are not yet describing the quantum domain which PTI seeks to describe. Instead, they are describing the realization of a particular classical state from an idealized, abstract set of equally possible states, where the realization can be attributed to the existence of a quantum domain that can "precipitate" a particular classical state by way of random quantum fluctuations. One can therefore point to the fluctuation precipitating the specific outcome as the "asymmetrical cause" required by Curie's principle.

If we return to the case of spontaneous symmetry breaking in the Standard Model, clearly we are dealing with symmetry breaking in a purely quantum context: the system comprises the vacuum and the Higgs field, purely quantum entities. If we want to try to follow the same procedure and to seek a specific cause – however fleeting and random – for the choice of one of the infinite set of possible vacuum states, we either have to suppose that it also stems from fundamentally indeterministic quantum fluctuations of the vacuum, or postulate fluctuations in some deeper realm that lies outside any current theory. The point is still that "the actual breaks the symmetry of the potential," however this is accomplished. The only alternative is to postulate that SSB in the Standard Model requires a "many worlds" interpretation, in which SSB gives rise to many possible worlds, each with a different vacuum state. But this is certainly not the usual approach, which simply assumes that the actual universe corresponds to one particular vacuum state.

The foregoing account of spontaneous symmetry breaking can be consistently extended to PTI's account of the realized notion of one particular transaction out of several, or even many, incipient ones. The mathematics describing the situation provides us with a set of possible states of the system, but only one of those can be realized. The new feature appearing in the PTI account is that this set of possible outcomes is weighted by the square of the probability amplitude for that outcome. So the proposed interpretation extends the basic principle of spontaneous symmetry breaking into a weighted type of symmetry breaking over a set of possible states: they certainly cannot all be realized (just as in the case of classical symmetry breaking), so the natural interpretation of the weight of a possible state is as a physical propensity, corresponding to an objective probability of the actualization of the state in question. If we like, we can call this a "weighted symmetry breaking" or WSB.

Again, note that the establishment of a set of incipient transactions does not require us to adopt a "many worlds" interpretation, any more than the "Mexican hat" potential establishing an infinite set of possible ground states requires an "infinitely many worlds" interpretation in the Standard Model. We simply infer that one of the set of possibilities is actualized; Hilbert space describes possibilities and their interactions, while spacetime is the arena of actualized transactions.

4.3.3 Symmetry breaking creates structure

The authors of *Fearful Symmetry* further note that instances of symmetry breaking give rise to concrete structures that seem to reflect design or intent. An interesting example is found in crop circles. The authors point out that an unblemished field of corn has a very high degree of symmetry: translation, reflection, and rotational symmetries. If symmetry is broken at a point – say by a falling object, such as a hailstone – the effect will ripple out radially to create a circle. A particular rotational center has been chosen from among an infinite number of equally "eligible" ones, and structure is born. The structure must obey the underlying rotational symmetry and it is also constrained by the nature of the objects comprising it. In the case of the crop circle, the physical properties of the cornstalks, together with the energy of the falling object or other precipitating event, will dictate the radius of the circle.

The crop circle provides an analogy for the symmetry breaking of PTI in the following way. Let us think of the structured phenomena appearing in the cornfield (not the field itself, but its crop circles) as the spacetime manifold. The field itself can be thought of as a kind of "offer wave." An offer wave is, in general, a very symmetrical object – consider, for example, the spherically radiating source of the Renninger experiment discussed in Cramer (1986). This experiment depicts such a source enclosed by a spherical absorbing surface of radius r; i.e., a collection of absorbing atoms, each of which is located at a distance r from the source. According to TI, every atom responds to the offer wave component received by it (attenuated by a factor of $1/r$) with a confirmation wave. The latter can be thought of as analogous to the "hailstone" which creates a crop circle; but in this case, we have a symmetrical distribution of "hailstones" as well, each of which *could* create its own circle.

Now, suppose all those hailstones *really* fell and each one created its own crop circle. The result would be a chaotic mess of overlapping cornstalks – structure would be lost. That is, there would be no more structure than before the hailstorm.[19] If the structured phenomena comprise the spacetime manifold, clearly not all the potential "crop circles" can be realized. Thus, nature chooses only one incipient transaction (crop circle) for any given interaction between offer wave and set of absorbers. The structure that results is what we experience as a phenomenon in spacetime. As noted above, the structure reflects the properties of the medium – which is not some spacetime substratum but rather the set of possibilities presented by offer and confirmation waves. It is in this sense that the latter are real: they dictate and constrain specific qualities of the actualized structures, just as the physical properties of the cornstalks and the energy of the incoming hailstone dictate the radius of the crop circle.

Nevertheless, the reader might protest: "So why is this not a 'many worlds' theory? Couldn't we say that all the circles really happen, each in a separate world?" There is nothing to stop someone from taking this view, but again, that is not what

[19] Of course, the analogy is not perfect, because (to push the analogy even further) one could always view the resulting effect as having more structure than a blank canvas. I assume here that the "circles" correspond to the highly structured spacetime phenomena observed in the real world which enable it to be described by mathematically structured theories in a way that the chaotic, "all-hailstones" cornfield could not be.

is routinely done in the context of other types of spontaneous symmetry breaking. Instead, it is supposed that vacuum fluctuations can trigger a particular solution from among the infinite possible ones. Moreover, for consistency, if one chooses to impose a "many worlds" interpretation on the set of incipient transactions and assert that they all occur in separate worlds, then one must say the same for the set of possible vacuum states appearing in the Higgs *et al.* mechanism, as well as for the competing sets of possible trajectories for any classical chaotic system (e.g., the double pendulum). I think that a more economical and elegant interpretation is that, in the words of Stewart and Golubitsky, "the actual breaks the symmetry of the potential."

4.4 "Transaction" is not equivalent to "trajectory"

In PTI, a transaction is not equivalent to the establishment of a classical spacetime trajectory; that is, a determinate path from one spacetime point to another. For example, in the two-slit experiment discussed in Chapter 1, with both slits open, a transaction transferring momentum from the emitter to a particular absorber X on the final screen does not establish a particular spacetime path. It retains the wavelike characteristic of a non-localized phenomenon in that the quantum went through both slits (or, technically, did not really "go through" slits in spacetime at all; see Chapter 7). This feature is best understood in the Feynman "sum-over-paths" approach to propagation in quantum mechanics, which I will now briefly review.

4.4.1 Review: Feynman "sum over paths"

Non-relativistic quantum theory is usually formulated in terms of Schrödinger's equation for the propagation of a "wave function," $\Psi(x)$, which is a particular solution to this equation. The wave function is a type of probability amplitude as discussed in Chapter 1 (specifically, an amplitude for a quantum to be found at position x if it has been prepared in the state $|\Psi\rangle$; technically, $\Psi(x) = \langle x|\Psi\rangle$). Richard Feynman formulated another approach to this probability amplitude (applicable also in the relativistic regime) by imagining a quantum as an entity that gets from one point to another by taking all possible spacetime paths (thus reflecting its "spread-out," wavelike nature).[20] While PTI considers the basic ontological quantum entity to be described by the state vector $|\psi\rangle$ rather than by a wave function which is a projection of the state vector onto the position basis, we can gain insight into the relationship of transactions to spacetime trajectories by considering Feynman's approach.

[20] For an eminently readable and delightful introduction to this formulation, the reader is encouraged to consult Feynman's popular book *QED: The Strange Theory of Light and Matter* (1985).

The Feynman sum-over-paths method asks the question: what are all the possible paths that a hypothetical particle could take from point A to a final point B? (We can think of A and B as spacetime points in a heuristic sense, but we should not assume that the particle "really" takes all paths as trajectories in a pre-existing spacetime substance – remember that these are just *possible* paths, in the submerged-part-of-the-iceberg sense.) One then adds up all the possible paths in a particular way (reflecting that they have both magnitude and phase), giving what can be called the "Feynman amplitude" for getting from A to B. If there are no obstacles (i.e., absorbers) of any kind between A and B, it turns out that the path predicted by this procedure is the ordinary classical path between A and B – that is, the path that a baseball would follow. This path can be considered a classical trajectory because there is virtually no uncertainty about it: one can predict with an extremely high degree of precision where the object will be at any given time as it propagates from A to B. In fact, this "sum-over-paths" process is an application of the "principle of least action" (PLA), also sometimes known as Hamilton's principle, after William Hamilton who formulated it. It says that nature chooses the path between two endpoints A and B for which the action (a quantity related to the difference between an object's kinetic and potential energies) is a minimum. (It turns out that such universal laws as Newton's laws of motion and the laws of electromagnetism are derivable from this principle, so it is very powerful.)

The situation becomes more complicated (and interesting) when there are obstacles present, such as in the two-slit experiment discussed in Chapter 1. (The reader is referred to Feynman and Hibbs, 1965, section 1–4 for a detailed discussion of the path integral in the presence of various obstacles.[21]) The type of phenomenon that results depends on the nature of the system under study and the obstacles: specifically, whether its quantum wavelength (recall Chapter 1) is significant compared to the obstacles (and/or their separation). If that wavelength is significant, then we have a situation in which the single classical trajectory discussed above is replaced by several (or many) possible paths, with interference between them (that is, there is no clearly defined trajectory).

Thus, in the limit of very small wavelengths (which applies to ordinary macroscopic objects like baseballs), we regain the appearance of classical trajectories.

[21] Feynman makes an interesting comment in section 1–3 of Feynman and Hibbs (1965) regarding his formulation of the calculation necessary to obtain the probability of an event. In distinguishing between observable and unobservable alternatives for a particle (where its path through one or the other slit falls into the "unobservable" category), he apparently wants to deny the following type of description: "When you watch, you find that it goes through either one or the other hole; but if you are not looking, you cannot say that it goes either one way or the other!" Yet his alternative description, in terms of formulated rules for calculating the probabilities, essentially boils down to the situation in the quoted sentence. Those rules simply substitute the presence or absence of a measuring apparatus for someone "looking" or "not looking."

In the limit of very large wavelengths (applying to quantum objects like electrons), we get interference and no particular spacetime path or trajectory. Instead, we have probabilities for an electron to be found on any of the possible paths (in TI terms, for a transaction detecting a photon associated with a particular path), but uncertainty about which one it takes; in some sense (but not in the usual sense of a spacetime journey), it takes *all* those interfering paths.

How do transactions fit into this picture? In the small-wavelength limit, which corresponds to a macroscopic (readily observable) particle, an offer wave with a particular initial momentum $|p\rangle$ has a negligible amplitude to interact with any absorbers other than those defining a classical trajectory.[22] The only uncertainty about its "path" arises from a non-quantum randomness in its original momentum (i.e., epistemic uncertainty over its classical initial conditions). So the corresponding Feynman amplitude is simply that of the classical path between A and B which goes through whichever slit is available to the particle based on its (epistemically uncertain) initial momentum. Similarly, any confirmation wave corresponds to the same slit as its prompting offer wave, and we have a "single-slit" transaction. Thus, if we want the total probability that the particle lands at B, we just add the separate probabilities for the classical trajectories through each slit, corresponding to the two distinct incipient transactions, and there is no interference. We can speak of the particles as pursuing a particular trajectory, but we just don't know which one (unless we add a detector at the slits). Thus, epistemic uncertainty, which requires that we add probabilities instead of amplitudes, is interpreted in PTI as simple uncertainty over which transaction occurred.

In the quantum (large-wavelength) limit applying to particles such as electrons, the offer wave has a significant amplitude to interact with absorbers defining the boundaries of both slits. In Feynman's terms, there are two basic ways for the particle to get from the source A to a final point B on the screen (each corresponding to many individual paths): one going through slit 1 and the other going through slit 2. But, unlike in the small-wavelength limit discussed above, all possible paths through both slits are superimposed in the offer wave reaching B, and similarly for the confirmation wave returned from B. (The latter point is particularly relevant for the two-slit experiment, addressed below.) The particle, in "going from A to B," does not follow a single clearly defined path. Given its emission from a source

[22] A basic postulate of TI is that offer waves can only interact with absorbers representing a matching state, as indicated in equation (4.1), in which only the component $|A_i\rangle$ of the original offer wave prompts a confirmation from absorber A_i. For small wavelengths and initial momentum corresponding to the absorbers defining the boundary of slit 1, the offer wave has essentially zero amplitude to interact with any of the absorbers defining slit 2 (that is, the boundaries of the slit). If one is concerned about a possible implicit reference to spacetime configurations, it should be kept in mind that spatiotemporal indices index points on fields, not spacetime points, in keeping with the non-substantivalism espoused in this work. Further details are found in Chapters 7 and 8.

at A, there is an amplitude for the particle to be detected at point B, but the corresponding transaction does not correspond to a well-defined spacetime trajectory; it merely transfers energy from point A to point B. Again, we see that the term "particle" is misleading, because it is *not* the case that a small, localized corpuscle propagated through space. Instead, a quantum of energy (and possibly other conserved quantities) was subtracted from the source at A and added to the absorber at B.

Thus, transactions can be considered simply a version of the Feynman sum-over-paths approach, with the added feature that absorption processes generate confirmations which in turn give rise to weighted incipient transactions. Taking into account the confirmation at B requires us to multiply the Feynman (retarded, or future-directed) amplitude for A $\rightarrow$ B by the advanced amplitude B $\rightarrow$ A, yielding the Born Rule. Just as in the Feynman amplitude for a quantum to get from A to B in the two-slit experiment, there is no well-defined spacetime trajectory. A trajectory exists only in an idealized classical (zero-wavelength) limit. While one can define the "amplitude of a path" for a quantum particle in the context of the Feynman picture, this does not correspond to a well-behaved probability in the absence of a sequence of actualized transactions defining the associated trajectory. This point is elaborated in Appendix B.

4.4.2 "Trajectories in a bubble chamber"

Transactions do not necessarily establish specific trajectories, so how can PTI account for the "trajectory" appearing when a subatomic particle propagates through a bubble chamber? A bubble chamber track is created by the ionization of molecules in the medium, which then act as catalysts for the formation of bubbles. What we actually see is a chain of bubbles forming around a chain of ionized molecules. In a typical bubble chamber interaction, a highly energetic quantum enters the chamber and is scattered by the first interacting molecule (in the process ionizing the molecule), but is not annihilated. Instead, it loses energy to the ionization process and continues on to subsequent molecules, repeating the process until all its energy has been "bled off."

Standard theoretical approaches to the passage of energetic charged particles through a medium use either the Bethe–Bloch equation (cf. Bethe, 1930) or the alternative Allison–Cobb (AC) approach (Allison and Cobb, 1980). The latter models the incoming particle as being surrounded by a cloud of virtual photons interacting with a dielectric medium (i.e., the atoms/molecules are polarized due to electromagnetic interactions). Classically, the photons are doing work against the field due to the polarized medium. Quantum mechanically, what is being calculated is the probability of the energy transfer by virtual photons of energy $E = \hbar\omega$.[23] The theory works by modeling the virtual photons (rest mass $\neq 0$) as real photons (rest mass $= 0$), so it is not exact, but it is empirically well-corroborated.[24] It treats each interaction of the virtual photon cloud with a target gas molecule as an independent scattering event.[25]

The AC model lends itself to a transactional interpretation if we consider each scattering event as the exchange of OW and CW between the incoming particle and the target gas molecule. In effect, the incoming particle acts as an emitter of OW and the gas molecule acts as an absorber (with the transferred energy being used to ionize the molecule). The probability of energy transfer for each scattering event (the square of the amplitude associated with each event) is simply the probability of a transaction; thus, the rate of energy loss is the rate of transactions transferring energy from the incoming particle to various gas molecules. The result is a chain of ionized molecules whose character reflects the specific properties of the incoming particle and the medium. The chain will be appropriately curved in the presence of a magnetic field, since the scattering computation takes into account whatever electromagnetic field is present in the medium. Thus we get the appearance of a "trajectory," which results from the ionization of gas molecules due to transfers of energy, via transactions, from the incoming quantum. But we should not let this mislead us into thinking that the incoming particle pursued a well-defined spacetime trajectory in the absence of the bubble chamber absorbers. I elaborate on this and related metaphysical points in Chapter 7.

4.5 Revisiting the two-slit experiment

To conclude this chapter, I revisit the basic two-slit experiment in the context of the Feynman sum-over-paths picture. In the previous chapter, I considered Feynman's account of a generalized two-slit experiment that allows for a measurement, of varying degrees of sharpness, of the electron's "path" through the slits. Here I consider only two possibilities: (i) no "which-slit" measurement at all or (ii) a very sharp "which-slit" measurement. However, as noted in the previous section, the electron doesn't really pursue a "path"; results of measurements reflect transactions between an emitter and one or more absorbers, based on amplitudes of interaction between OW, CW, and potential absorbers, and do not imply or "reveal" a spacetime trajectory.

[23] One must integrate over all possible energy and momenta independently, since these are virtual photons.

[24] The cross-section for scattering of a virtual photon must be approximated by known cross-sections for real photons.

[25] The approximation consists primarily in the fact that well-defined (in the sense of theoretically exact) scattering events only apply to real particles as the "in" and "out" states, whereas these "in" and "out" states are virtual particles, usually represented only by internal lines.

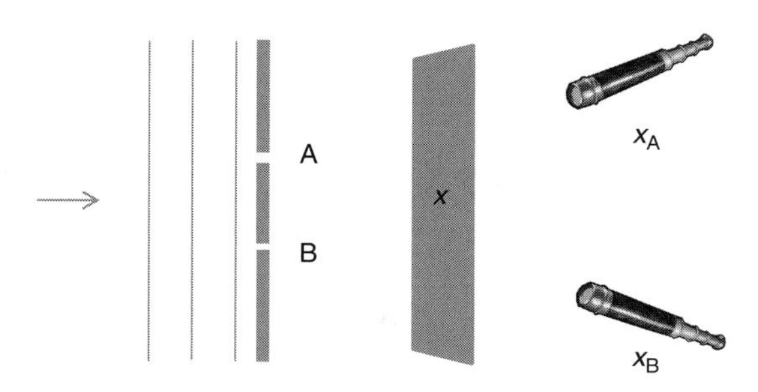

Figure 4.10 Two-slit experiment with an optional "which-slit" measurement via telescopes.

Returning to the basic two-slit setup (see Figure 4.10), the final detection screen can be considered as being composed of a large number of absorbers, each corresponding to all the possible positions x on the screen. Recalling the discussion in the previous section, we see that for large wavelengths and in the absence of a "which-slit" measurement, the OW has a significant amplitude to interact with the absorbers defining the boundaries of both slits A and B, as does the CW generated by any of the absorbers in the final detection screen. That is, a CW is generated by each absorber x and all such CW must be considered as having access to both slits (as opposed to only one or the other slit). The probability of detection at a particular position x is the product of the OW amplitude at x (which is the OW component reaching x as opposed to some other position x') and the CW amplitude generated at x and terminating (by way of both slits) at the emitter. In more quantitative terms, the amplitude of the OW reaching a position x can be represented as $\langle x|\Psi\rangle = 1/\sqrt{2}[\langle x|\mathrm{A}\rangle + \langle x|\mathrm{B}\rangle]$ and the amplitude of the CW generated at x as $\langle\Psi|x\rangle = 1/\sqrt{2}[\langle\mathrm{A}|x\rangle + \langle\mathrm{B}|x\rangle]$.

The probability given by the product of the amplitudes is therefore (as in the TI version of the Born Rule including the projection onto $|x\rangle$)

$$P(x)|x\rangle\langle x| = \langle x|\Psi\rangle\langle\Psi|x\rangle|x\rangle\langle x| = 1/2[\langle x|\mathrm{A}\rangle + \langle x|\mathrm{B}\rangle][\langle\mathrm{A}|x\rangle + \langle\mathrm{B}|x\rangle]|x\rangle\langle x|$$

$$= [\langle x|\mathrm{A}\rangle\langle\mathrm{A}|x\rangle + \langle x|\mathrm{B}\rangle\langle\mathrm{B}|x\rangle + \langle x|\mathrm{A}\rangle\langle\mathrm{B}|x\rangle + \langle x|\mathrm{B}\rangle\langle\mathrm{A}|x\rangle]|x\rangle\langle x| \quad (4.3)$$

where the last two terms reflect interference between the slits.

Now suppose we consider a "which-slit" measurement of the kind envisioned in Wheeler's famous "delayed choice" experiment. This consists of replacing the final detection screen with a pair of telescopes, each focused on one of the slits A and B. (This version is done with photons.) What the focusing mechanism does, in terms of the (time-reversed) Feynman paths picture, is to greatly increase the amplitude for a

CW to interact with the absorbers defining the boundaries of the slit at which the telescope is aimed, while making the amplitude for interaction with the other slit negligible. This means that, even though the OW has a finite amplitude to interact with both slits, the CW generated by either telescope does not.[26] (One can also see this as the OW component corresponding to slit A having negligible amplitude to reach telescope B and vice versa.) Again in more quantitative terms, the probability of detection at a particular telescope x_A must be specified in the absorber basis and is given by

$$P(x_A)|x_A\rangle\langle x_A| = \langle x_A|\Psi\rangle\langle\Psi|x_A\rangle|x_A\rangle\langle x_A|$$

$$= 1/2[\langle x_A|A\rangle\langle A|x_A\rangle + \langle x_A|B\rangle\langle B|x_A\rangle]|x_A\rangle\langle x_A| \quad (4.4)$$

which exhibits no interference since $\langle x_A|B\rangle$ is zero. The point is that these probabilities are reflections of physical amplitudes of interactions between emitters and absorbers, and do not indicate spacetime trajectories. No well-defined particle trajectory can be inferred based on amplitudes which apply to the pre-spacetime level. For example, the OW does have a finite amplitude to interact with either slit boundary (i.e., both $\langle\Psi|A\rangle$ and $\langle\Psi|B\rangle$ are different from zero), so one can think of the OW as "having gone through" both slits, but since the CWs do not (i.e., $\langle B|x_A\rangle = \langle A|x_B\rangle = 0$), the transactions available do not exhibit two-slit interference. The "particle" is no more and no less than whatever transaction is actualized.

In the next chapter, I consider some specific challenges to TI to see how they are handled under PTI, as well as some applications of PTI to various thought-experiments such as "interaction-free measurements" and the "quantum eraser."

[26] Technically, the "absorber" for each telescope is a macroscopic object comprising many microscopic absorbers (recall Section 3.4). Here the phrase "telescope at x_A" just means the entire class of microscopic absorbers corresponding to telescope A.

5

Challenges, replies, and applications

5.1 Challenges to TI

Tim Maudlin considered TI in his book *Quantum Nonlocality and Relativity* (2002, pp. 199–201), which explored the apparent tension between quantum theory and relativity in terms of non-local effects and influences. He concluded at that time that TI was not viable based on a type of "thought-experiment" which seemed to imply an inconsistency. Maudlin's challenge and similar challenges have been addressed by several authors, who have argued that it is not fatal for TI.[1] The present author is amongst those who have argued that Maudlin-type challenges are not fatal, but the basic concern behind them is an important one that has prompted further development of the interpretation. A key component of this development of "possibilist TI" or PTI is that offer and confirmation waves are physical possibilities which are sub-empirical and pre-spatiotemporal. Another component is the necessity to embrace a "becoming" view of events rather than a "block world" view. The latter will be more fully examined in Chapter 8, but I introduce it here in Section 5.1.3 as part of the argument concerning a similar challenge posed to standard quantum mechanics by the "delayed choice experiment" proposed by John Wheeler.

As discussed in the previous chapter, OW and CW should not be thought of as propagating within spacetime (in either temporal direction); but rather as acting instead at a pre-spacetime (PST) level. Actual spacetime events are emergent from the transactional process; they are supervenient on that process rather than being present *a priori* as part of a spacetime substance or "block world," as is assumed in Maudlin-type challenges. While the PTI ontology – especially the sub-empirical, extra-spatiotemporal nature of the offers and confirmations – has been viewed with some initial skepticism, it should be kept in mind that most

[1] Berkovitz (2002); Cramer (2005); Kastner (2006); Marchildon (2006).

competing interpretations incorporate sub-empirical features as well. For example, the MWI assumes a sub-empirical, extra-spatiotemporal splitting of worlds or observers, and the Bohm theory assumes a sub-empirical, extra-spatiotemporal "guiding wave" which is conceptually very similar to TI's offer wave. Because the Hilbert space structure of the theory is not reducible to that of spacetime – the manifold of empirical events – any realist interpretation of quantum theory must acknowledge that the mathematical formalism refers (at least in part) to something transcending the empirical realm.[2] This inevitable message of the theory is again reflected in Bohr's comment that quantum processes "transcend the spacetime construct."

5.1.1 The Maudlin challenge

The Maudlin challenge is a critique of the "pseudotime" account presented in Cramer (1986), in which transactions are established in a forward-and-backward temporal process between an emitter and a set of absorbers. It proposes a thought-experiment in which the placement of a distant absorber for a possible transaction is contingent on the failure of a competing transaction with a nearby absorber, so that one cannot think of the absorbers as a static backdrop for the "competition" among incipient transactions as assumed in Cramer (1986).

The basic argument can be summarized as in Figure 5.1. A source emits massive (and therefore, Maudlin assumes, slow-moving)[3] particles either to the left or right, in the state $|\Psi\rangle = \frac{1}{\sqrt{2}}[|R\rangle + |L\rangle]$, a superposition of "rightward"- and "leftward"-propagating states. OW components corresponding to right and left are emitted in both directions, but, in this arrangement, only detector A can initially return a CW (since B is blocked by A). If the particle is not detected at A (meaning that the rightward transaction failed), a light signal is immediately sent to detector B, causing it to swing quickly around to intercept the particle on the left. B is then able to return a CW, but it is only of amplitude $\frac{1}{\sqrt{2}}$ and yet the particle is certain to be detected there, which Maudlin claims is evidence of inconsistency on the part of TI. He also argues that the "pseudotime" picture cannot account for this experiment, since the outcome of the incipient transaction

[2] An interesting image reflecting this mathematical fact can be found on the cover of Bub's *Interpreting the Quantum World* (1997). The cover image shows M. C. Escher's famous print "Waterfall," depicting a scene with a physically impossible topology (i.e., one that could not actually fit into spacetime). Three separate areas of the print are highlighted, and each of these could exist in isolation in spacetime, but the global connections between them cannot. In the interpretation proposed herein, the smaller highlighted "normal" areas represent the actualized transactions, and the larger shaded area of topologically "impossible" global interconnections belong to the pre-spacetime realm of possibility (i.e., offer and confirmation waves).

[3] Note that this argument assumes that the particle's offer wave propagates in spacetime at a speed less than c. However, one can question whether offer waves propagate at the speed of their associated quanta. This issue is explored in Chapter 8.

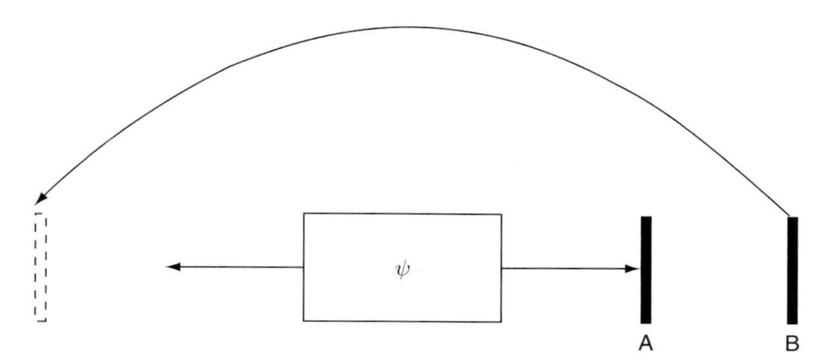

Figure 5.1 The Maudlin contingent absorber experiment.

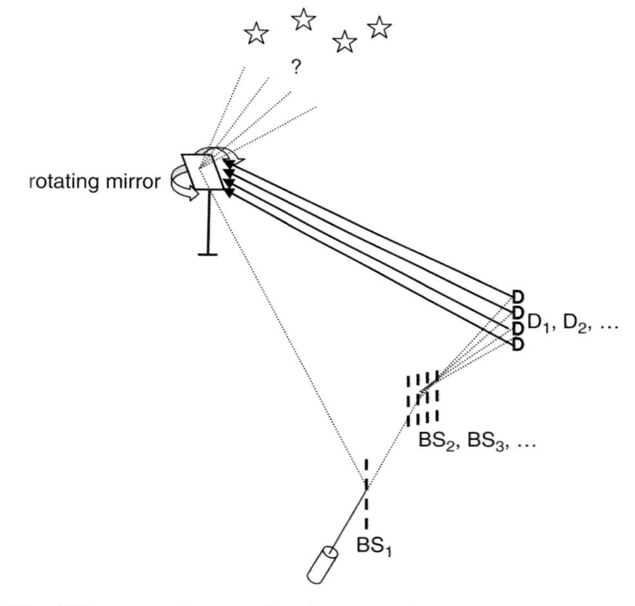

Figure 5.2 The Elitzur contingent absorber experiment.

between the emitter and the nearby absorber must be decided without a CW from the more distant absorber. This undercuts the original "pseudotime" account, which assumes that all CW are received by the emitter at once and the choice of which transaction is realized is made based on a sort of "competition" between incipient transactions.

There are quite a few variations on the original Maudlin challenge. Elitzur (private communication, 2009) has proposed a variation (see Figure 5.2)[4] in

[4] Figure 5.2 is Elitzur's illustration. Used with permission.

which the choice of a distant object from among several possible ones is determined by whether a nearby detection occurs. The part of the offer wave heading to the distant region is reflected to outer space by a rotating mirror, the angle of which is set by the nearby detector's response. The basic scenario can be elaborated by splitting the nearby component as often as one likes, thus sharpening the causal loop aspect: if none of the nearby detectors click, then it becomes certain that the "default" target will participate in the actualized transaction. Yet according to the account in Cramer (1986), that object's confirmation wave supposedly participated in the "transaction roulette" in which the closer non-detections ended up selecting it. An additional variation was presented by Berkovitz (2008). He proposed an experiment involving an EPR-type entangled state of two particles, in which the choice of measurement made on one of the correlated particles is contingent on the outcome of a given measurement on its companion, which is measured first.

Miller (2011) has proposed a challenge involving photons. In this version of the challenge, a photon is split by a half-silvered mirror into two beams A and B; the beam in B is temporarily detoured by a fixed set of mirrors to delay its absorption by absorber B. If it is not detected at A at $t = 1$, a moveable mirror is quickly inserted into the beam going to detector B such that the OW component in that arm is diverted to detector B′ (perhaps with different properties such as a polarization filter). This makes the specific CW component corresponding to arm B, and returned to the emitter, dependent on the outcome at $t = 1$. Thus there appears to be "no fact of the matter" about which CW are present, B or B′.

It was argued in Kastner (2006) that the consistency component of the Maudlin challenge is far from fatal. Specifically, there is always an offer wave (OW) received by the nearby fixed absorber, with a responding confirmation wave (CW). This overlap of OW and CW is an incipient transaction, existing in all runs of the experiment, which has a probability of ½ of being actualized. If it fails, then the transaction with the distant absorber must be actualized. The CW from the distant absorber, of amplitude $\frac{1}{\sqrt{2}}$, multiplied by the OW component of amplitude $\frac{1}{\sqrt{2}}$ which generates it, reflects the fact that this transaction only occurs 50% of the time. Thus TI can indeed provide a consistent account of the Maudlin experiment, although the original heuristic "pseudotime" description of Cramer (1986) cannot be applied. (I provide a slightly more technical account of how TI can provide a consistent set of probabilities for Maudlin-type experiments in the next section.)

Another response was given in Cramer (2005), which proposed that transactions can form in a hierarchical manner with respect to the spacetime intervals involved, with smaller intervals taking ontological precedence: thus a transaction with a distant absorber can come into play only if the competing transaction with a nearby absorber fails. Miller has argued, with reference to his thought-experiment above, that this approach cannot be applied for photons, since all photon spacetime intervals are zero. Cramer has argued that the mirror-bounced path of the delayed photon OW component constitutes a timelike interval and therefore his hierarchy approach can still be applied in this case.[5] However, the hierarchy approach to resolving these "causal loop" issues would appear to require that the incipient transactions have clearly defined spacetime intervals, which may not be the case in certain types of interaction-free measurements to be described below.[6] I believe that the correct approach to resolving these issues is the one presented in the next section.

5.1.2 Contingent absorber experiments and the delayed choice experiment[7]

Maudlin-type challenges can be called "contingent absorber experiments" (CAE), since they involve placement of absorbers, or a particular choice of measurement, only contingent on a previous outcome. In this section, I point out that CAE cannot be considered refutations of TI unless the delayed choice experiment (DCE) is also considered a refutation of standard quantum mechanics. This section is rather technical, so readers interested in a more qualitative overview can skip to Section 5.1.3.

Maudlin's challenge has two distinct features: (1) it seems to involve a situation not amenable to the usual "echoing" picture as given in Cramer (1986) – in particular, there seems to be no definite account of what CW are present; and (2) it is not clear that the probabilities are consistent or well-defined. Regarding (1), the problem, according to the usual way of thinking, seems to be the following. At $t = 0$ the OW is emitted, but since CW propagate in the reverse temporal direction from absorbers to the emitter, whatever CW are generated at a later time must "already" be back at the emitter. Berkovitz (2002) explicitly augments the emitter with a label corresponding to the presence (or absence) of a CW (and the particular state of the CW), calling this the "state of the emitter."[8] Thus, depending on the outcome, there are two different "states of the emitter," so there seems to be no "fact of the matter" about which one is the "correct" one – in contrast to the standard case, discussed in Cramer (1986), in which all absorbers are present at the initial OW. As alluded to earlier regarding feature (2), Maudlin argues that once A has failed to detect the

[5] Private communication (2011).

[6] There may be a further difficulty with the hierarchy approach. Cramer has argued (in a private communication) that the hierarchy only comes into play for causal loop-type situations and that in general, the incipient transaction with the shorter spacetime interval does not have any ontological priority over the longer one. However, it is unclear, at least to this author, why nature would proceed in this fashion only for causal loop situations. This seems to give the hierarchy account an ad hoc property.

[7] This subsection and the next are based on Kastner (2011b).

[8] TI doesn't adopt this ontology, since an emitter is not described by a particular CW, but I understand Berkovitz to be attempting to be precise about the circumstances surrounding CW generation, so I'll use his terminology in this discussion.

particle, it is certain to be detected at B, implying that the B outcome should have a probability of 1; but the OW/CW incipient transaction corresponding to detection at B has a weight of only ½ (this corresponds to the Born Rule of standard quantum mechanics and is simply obtained in TI as the product of the amplitudes of the OW and CW components). He therefore argues that TI's probabilities, given by the weights of the associated incipient transactions, seem inconsistent.

The short response to objection (2) is that the weights of incipient transactions are *physical* in nature rather than epistemic (i.e., they are not based on knowledge or ignorance of an observer). In Maudlin's experiment, detection at B will only occur in ½ of the trials, and that is what the weight ½ describes; not the observer's knowledge, based on failure of detection at A in any particular trial, that the particle will be detected at B. I return to (2) in a little more detail below.

Objection (1) is the issue more likely to be seen as problematic for TI: it seems to thwart the idea of a clearly defined competing set of incipient transactions as described in Cramer (1986), since one or more absorbers are not available to the emitter unless another outcome occurs (or fails to occur). This seems to set up causal loops, in Berkovitz's terminology, as follows. At $t=0$ the OW is emitted; at $t=1$ a CW is returned from A, so the "state of the emitter" can be represented as OW(A). Now suppose the transaction between the emitter and A fails. Then a signal is sent from A to swing B into position to intercept the OW on the left at $t=2$. Absorber B now returns a CW, so the "state of the emitter" at $t=0$ is OW (A,B). But here's the causal loop: if the state of the emitter is OW(A,B), it is already certain at $t=0$ that *B* is in place and the particle must be detected on the left. On the other hand, if the state of the emitter is OW(A), then it is certain at $t=0$ that the particle must be detected on the right. We seem to have two causal loops that contradict each other: both seem "predestined" as of $t=0$, but they obviously can't both happen.[9]

Before considering the resolution of these (1)-type issues, let's return to objection (2) to see in more detail how TI's probabilities are indeed well-defined in CAE. In the Maudlin experiment, if there is really no other absorber for the OW component heading toward the left, theoretically there may be no incipient transaction on the left, but we may still define the relevant probabilities by taking

[9] But see Kastner (2006), which argues that the probability of ½ applies to each loop, and that the emitter state (which is the one that seems to be self-contradictory) should not be viewed as the "branch point" between loops, but rather the incipient transaction between the emitter and the fixed absorber which then determines which of the possible "emitter states" is actualized. This implies that the past need not be viewed as determinate, as is also proposed here. An alternative solution by Marchildon (2006) argues that CW are well-defined in such experiments in a "block world" picture, which in our view amounts to a "hidden variable" approach; i.e., that there is always a fact of the matter about the "state of the emitter" but it is unknown to experimenters. Since I take quantum mechanics as complete, I view the status of quantum objects as genuinely indeterminate.

the total sample space as consisting of the outcomes (yes, no) for the question "Is the particle detected on the right"? That is, actualization of the incipient transaction between the emitter and A is the answer "yes" and its failure is the answer "no." Each answer's probability is ½. This is a natural step of applying the law of the excluded middle to cases in which there is only one incipient transaction corresponding to a given outcome for a particular observable: the only physical possibilities are that the one possible transaction either succeeds or that it does not.

For the more general case, note that we can do a "spectral decomposition" of any observable, i.e., we can express it in terms of a sum of the projectors onto its eigenstates. In TI these mathematical projectors represent incipient transactions, while the density operator (a weighted sum of projection operators) corresponds physically to the set of weighted transactions resulting from the encounter of an emitted OW with the absorbers actually available to it.[10] The projectors constitute a complete disjoint set covering all possible outcomes, and defining a Boolean (classical) probability space. Suppose there are N projectors corresponding to N possible outcomes (each identified with an absorber). Label them each n ($n = 1, N$), where the weight of the nth incipient transaction, corresponding to the probability of its associated event being actualized, is $P(n)$.[11] Then $P(1) + P(2) + \ldots + P(N-1) + P(N) = 1$. If there are incipient transactions corresponding to N–1 of the outcomes but not for a particular outcome k, then the probability that none of the N–1 incipient transactions is actualized is $1 - \sum_{n \neq k} P(n) = P(k)$. This is the generalization of the two-outcome experiment discussed above, in which the probability of the answer "no" to the question "Is the particle on the right?" is the same as the probability of the answer "yes" to the question "Is the particle on the left"? Thus it is clear that the probabilities for various outcomes can be unambiguously defined for contingent absorber experiments or simply for experiments in which there is not a complete set of absorbers.

5.1.3 Delayed choice as a challenge for orthodox quantum mechanics

As promised, I return to the issue of an indefinite "state of the emitter" in the face of the different apparent causal loops presented by contingent absorber-type

[10] Specifically, in the expression $|x\rangle\langle x|$, "$|x\rangle$" represents the component of the OW absorbed by a detector corresponding to property x of observable X, and "$\langle x|$" represents the CW response of the absorber. If the OW and CW amplitudes are a_x and a^*_x, respectively, the associated weighted incipient transaction is represented by $a^*_x a_x |x\rangle\langle x|$. Note that the set of weighted transactions corresponds to von Neumann's "Process 1" or "choice of the observer as to what to measure," the physical origin of which remains mysterious in standard interpretations but which has an obvious physical interpretation in TI.

[11] The weight is the Born Rule, which in TI is simply the product of the amplitudes of the OW and CW comprising each incipient transaction; see the previous note.

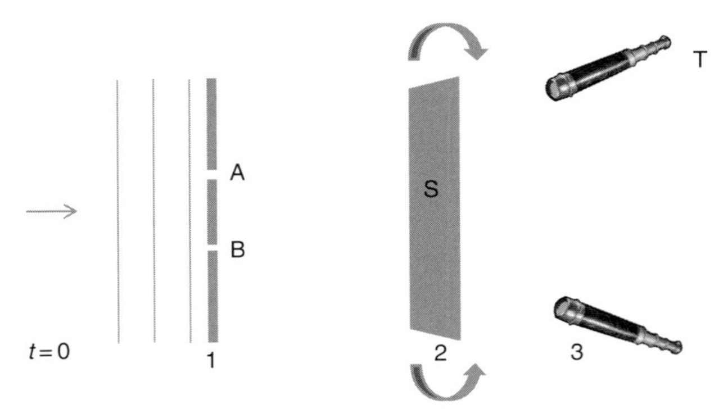

Figure 5.3 The delayed choice experiment.

experiments. The way to deal with this issue is by noticing that a similar conundrum already appears in standard, orthodox quantum theory in the delayed-choice experiment. Here is where I take note of the observation by Stapp (2011) that, even in orthodox QM, there is no "fact of the matter" about certain aspects of the past relative to the times of such delayed choices.[12] To see this, let us briefly review the DCE.

The standard (non-TI) presentation of the DCE is as follows (see Figure 5.3). (1) At $t=0$, a photon is emitted toward a barrier with two slits A and B. (2) At $t=1$, the photon passes the barrier (i.e., I discard runs in which the photon is blocked by the barrier). (3) The photon continues on to a screen S on which one would expect to record (at $t=2$) an interference pattern as individual photon detections accumulate. (4) However, the screen may be removed before the photon arrives (but after it has passed the slit barrier), revealing two telescopes focused on each slit. (5) If this happens, the two telescopes T will perform a "which-slit" measurement at $t=3$ (recall Section 4.5), and the photon will be detected at one or the other telescope, indicating that it went through the corresponding slit (i.e., there is no interference). The decision as to whether to remove S or not is made randomly by the experimenter.

Note that the photon has already passed the plane of the slits before the observer has decided whether to measure "which slit" or not. Thus, at a time $1 < t < 2$ prior to the observer's choice, *there is apparently no "fact of the matter" about the photon's state, including whether or not it has "interfered with itself."* The reader might object that the photon's state is simply a superposition of slits, $|\psi\rangle = (1/\sqrt{2})[|A\rangle + |B\rangle]$, but that actually assumes that no "which-slit"

[12] This work does not address the aspect of Stapp (2011), which considers the possibility of alteration of the statistical predictions of orthodox quantum theory. It deals only with Stapp's observations concerning standard quantum theory.

measurement takes place and implies that interference is present, which is uncertain given the experimental setup.

To see this fact clearly, we need to take into account situations involving both the preparation *and post-selection* of quantum systems. It is well known that a particle pre-selected in one state and post-selected in another state (where these can be states of two non-commuting observables) can be equally well described by *either* the pre- or post-selected state. This is because standard quantum mechanics gives a probability of unity for the outcome corresponding to either the pre- or post-selection state for a hypothetical measurement of either the pre- or post-selection observable conducted at a time between the two states.[13] A photon pre-selected in $|\psi\rangle$ as above, and post-selected via a "which-slit" measurement at $t=3$ (say yielding the outcome "slit A") can be described by either $|\psi\rangle$ or $|A\rangle$ at $t=1.5$; but a photon pre-selected in $|\psi\rangle$ for which the screen S is not removed *cannot be described at any time by a one-slit state*. Thus the photon's ontological status is undefined in a way that goes beyond the usual quantum indeterminacy – i.e., as exemplified by ordinary superpositions such as the state $|\psi\rangle$, or by ambiguous states based on a *specified* pre- and post-selection as discussed above and in note 6. In the case of the DCE, we have ambiguity not just based on a given pre- and post-selection but on *an uncertainty in the post-selection itself*, which translates into an essentially different kind of ambiguity in the ontological status of the photon between measurements.[14]

The reader may still think that the above indeterminacy of the photon state is just the usual (relatively benign) quantum indeterminacy. However, this is not the case; the delayed choice experiment also presents a "causal loop" problem for standard quantum mechanics, as follows. In the usual account, at $t=1$ a photon progresses past the slit plane.[15] If an experimenter later (at $t=2$) removes S to reveal a "which-slit" detector, this action means that the photon *must have only gone through one or the other slit at* $t=1$, since there can be no interference between paths corresponding to each slit.[16] According to the "block world" way of thinking,

[13] This is the standard time-symmetric ABL Rule (1964) for a measurement performed in between a pre- and post-selection. The inference that either the pre- or post-selected state could be attributed to a particle at an intervening time is a direct consequence of the calculus of probabilities applying to standard quantum theory and does not run foul of any illegitimate counterfactual usage of the ABL Rule. See Kastner (1999), equations (24)–(26) and supporting discussion, for why this is so.

[14] In the TI picture of the DCE, the photon's OW is perfectly well-defined; it is only its CW that is not well-defined. This is arguably a simpler way to understand the DCE.

[15] I say "the usual account" because in TI there is no "photon passing the slit plane." There is only an offer wave.

[16] Although it might be argued that the Copenhagen interpretation would not countenance *any* statement about the whereabouts of the photon prior to the choice, the usual approach to the DCE, and certainly Wheeler's approach, has been to infer that the choice of a "which-way" measurement determines what happens to the photon in the past. This is the whole point of Wheeler's amplification of the DCE to astronomical proportions as in his version with a photon wave function traveling from a distant galaxy and being split by gravitational lensing: to emphasize how present choices may affect an arbitrarily distant past.

this means (for logical and physical consistency) that the experimenter *must* place the "which-slit" detector and his choice never was free but predetermined, even though there was no causal mechanism forcing his choice.

Now, one might just conclude that this implies there is no genuine freedom of choice (a price gladly paid by many "block world" adherents), and the comparison between the two types of experiments (delayed choice vs. contingent absorber) ends there. However, we can sharpen the "causal loop" aspect of the delayed choice experiment by proposing that the experimenter does not choose but instead makes removal of S contingent on the outcome of a quantum coin flip; say, the outcome of the measurement of "up" or "down" of an electron in a Stern–Gerlach (SG) apparatus. That is, a SG measurement could be conducted independently but alongside the usual DCE, and a particular outcome occurring at $t = 1.5$, say "up," used to automatically trigger removal of S. Then the outcome of that "coin flip" *must also be predetermined* based on the effect of the delayed measurement which retroactively decides whether or not the photon "interfered with itself" while passing through the slits prior to that measurement. Yet quantum mechanics predicts that the outcome of the SG "coin flip" measurement is uncertain (e.g., has only a 50/50 chance). This is essentially the same alleged inconsistency presented for TI by CAE-type experiments, and which was a major concern of Maudlin (2002).

To summarize: in the TI case, the alleged inconsistencies presented by the CAE are (1) an apparent lack of "fact of the matter" concerning which CW are present at $t = 0$ and (2) the apparent causal loop paradox in which one or the other outcome must be predetermined, while the outcome of the incipient transaction in the fixed portion of the experiment is supposed to be uncertain, with a probability of ½ of "yes" or "no." In standard quantum mechanics and the DCE augmented by a quantum coin flip, the inconsistency concerns the apparent lack of "fact of the matter" about the ontological status of the photon, based on the uncertainty of its post-selection state, for times $1 < t < 2$; and the apparent causal loop paradox in which a particular measurement choice is mandated for the photon, vs. the prediction of 50/50 for the outcome of the quantum coin flip at $t = 1.5$ that determines the choice (removal of S or not). Note, however, that neither the ontological status of the photon prior to the choice nor the "state of the emitter" in the TI picture are empirically accessible, so there can be no violation of causality for either case in the form of a "bilking paradox" or other overt contradiction with experience.

Thus we see that the delayed choice experiment (especially when augmented to make the choice dependent on a quantum outcome) seems to raise the same sort of causal loop conundrum that has been used as a basis for criticism of TI. The point here is that *the possibility of an ontologically indeterminate situation* *at t that becomes determinate in virtue of a later outcome at $(t+\Delta t)$ is a feature of standard quantum mechanics itself* and therefore cannot be viewed as a defect of a particular interpretation such as TI. This should perhaps not be so surprising, since there is no contradiction with observation: the portions of the past that are indeterminate are empirically inaccessible. They become determinate based on events that select out of the possible past events certain actual ones, as discussed by Stapp (2011).

CAEs gain traction as apparent threats to TI based on the idea that there must be a "fact of the matter" about what CW exist *at the present time of the emission* – that is, when $t=0$ is "now" – and such experiments clearly make that impossible. But in fact there is no reason to demand this of TI, since there can likewise be no "fact of the matter" about the self-interference status of a photon when $t=1.5$ is "now" in the DCE in orthodox quantum mechanics.[17] (The two can be legitimately compared because CW in TI are no more empirically accessible than is the ontological status of an individual photon between measurements.) Furthermore, as shown above, a variant of the DCE raises the same types of consistency issues for standard quantum probabilities (in the usual implicit "block world" picture) as do CAE for TI. As emphasized by Wheeler, *the message of quantum mechanics is that the determinacy of certain aspects of the past depends on what happens in the present (i.e., the future of the past event(s) in question)*. Once one allows for this metaphysical possibility, the apparent inconsistencies vanish in both cases. The status of a photon's self-interference can be uncertain and contingent on a later measurement, just as the "state of the emitter" in TI can be uncertain and contingent on the outcome of a later incipient transaction.

5.2 Interaction-free measurements

Elitzur and Vaidman (1993) pioneered the idea of "interaction-free measurements" (IFM). Much has been written already about such experiments, and deservedly so, since they exhibit very clearly the counterintuitive, non-classical nature of quantum events. In Section 5.2.1 I discuss the original IFM; later sections discuss variations on this experiment, which present a challenge to TI in its original form.

[17] This locution obviously implies an "A-series" view of time. This aspect will be more fully addressed in Chapter 8. For purposes of this chapter, it may be noted that the "B-series" or 'block universe" view is part of the problem in that it implies causal loops that need not exist in an "A-series" picture. That is, the photon's status may be indeterminate at $t=1.5$ (now = 1.5) but determinate at $t=1.5$ (now = 3), where "now" is indexed by Stapp's "process time" (Stapp, 2011). The first t index can be thought of as the number of a row in a knitted fabric, while the second "now" index can be thought of as indicating which row of stitches is currently on the needle. There need not be any overt conflict between this picture and McTaggart's much-contested "proof" that time itself does not exist. We need only make a distinction between (i) the numbering of events with respect to each other and (ii) the inception of each event.

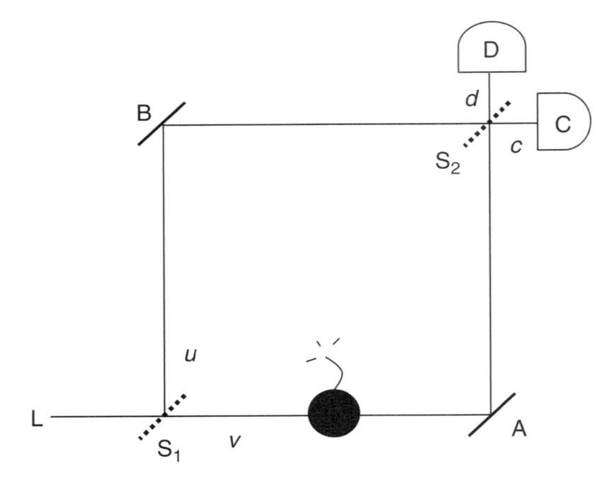

Figure 5.4 The Elitzur–Vaidman "bomb detection" interaction-free measurement.

5.2.1 The Elitzur–Vaidman bomb detection IFM

The original EV paper (1993) presents a way to examine a bomb to make sure it is working properly, but without activating (exploding) the bomb. Of course, the experiment is an idealization, but it provides a lovely illustration of the way a quantum system can "probe" its environment without necessarily interacting with it in a classical manner. The basic setup is illustrated schematically in Figure 5.4.

The laser L acts as a source of photons in a state we'll call $|s\rangle$. There are two beam splitters (half-silvered mirrors) S_1 and S_2, which transmit and reflect equal components of the incident state. Note that a photon described by a state such as $|v\rangle$ (corresponding to being found in arm v of the interferometer, see Figure 5.4) acquires a 90° phase change, corresponding to multiplication of a factor of i, upon reflection. (We disregard the total reflections at mirrors A and B because they don't affect the final result.) Thus, after two reflections, the state acquires a phase change of 180° and is multiplied by a factor of -1, etc.

The interferometer is set up so that a photon entering the device can only be detected at detector C. Considering just the empty interferometer with no obstruction in either of the arms u or v, this is accomplished as follows. Let's call the initial photon state from the laser source $|s\rangle$. Upon passing through the first beam splitter S_1, its state is transformed as

$$|s\rangle \rightarrow \frac{1}{\sqrt{2}}(i|u\rangle + |v\rangle) \tag{5.1}$$

that is, the initial state becomes a superposition of a transmitted component corresponding to arm v and a reflected component corresponding to arm u, with a phase shift factor of i as described above. (The factor of $\frac{1}{\sqrt{2}}$ indicates that these two components are equal in amplitude.)

Next, we have to consider what happens to each of the states $|u\rangle$ and $|v\rangle$ as they interact with the second beam splitter, S_2. Each of these states undergoes a splitting similar to that of the initial state $|s\rangle$, as follows:

$$|u\rangle \rightarrow \frac{1}{\sqrt{2}}(|c\rangle + i|d\rangle)$$

$$|v\rangle \rightarrow \frac{1}{\sqrt{2}}(i|c\rangle + |d\rangle) \qquad (5.2a, b)$$

If we substitute these expressions into the original state $|s\rangle$, we find that it evolves as follows:

$$|s\rangle \rightarrow \frac{1}{\sqrt{2}}(i|u\rangle + |v\rangle) \rightarrow \frac{1}{\sqrt{2}}\left[\frac{i}{\sqrt{2}}(|c\rangle + i|d\rangle) + \frac{1}{\sqrt{2}}(i|c\rangle + |d\rangle)\right]$$
$$= \frac{1}{2}[i|c\rangle - |d\rangle + i|c\rangle + |d\rangle] = i|c\rangle \qquad (5.3)$$

Thus, destructive interference between components corresponding to path $|d\rangle$ prevents the photon from reaching detector D and that detector will never activate; photons will always be detected at C. Thus, detector D is called a "silent detector" in this type of experiment. In technical terms, the probability of activation of detector C is given by the Born Rule, which prescribes that we square the projection of state (5.3) onto $|c\rangle$; thus we get

$$\text{Prob(C activated)} = -i \cdot i|\langle c|c\rangle|^2 = 1 \qquad (5.4)$$

Now, let's see what Elitzur and Vaidman have in mind as far as using this setup to examine a bomb without setting it off (see Figure 5.4). Keeping in mind the above analysis of the empty interferometer, consider the addition of an obstruction in arm v. We now have three possible experimental outcomes:

1. Detector C is activated.
2. Detector D is activated.
3. The photon is absorbed by the obstruction.

In this experiment, component $|v\rangle$ cannot reach S_2, so it cannot reach either detector. A photon described by $|v\rangle$ will inevitably be absorbed by the obstruction (outcome 3 above). The only component that has a chance of reaching the detector area is $|u\rangle$. Recalling that the original state $|s\rangle$ has equal components of $|u\rangle$ and $|v\rangle$ (5.1), the relevant probabilities are:

$$\text{Prob}(\text{C activated}) = \left|\langle c|\left(\frac{i}{\sqrt{2}}|u\rangle\right)\right|^2 = \frac{1}{4} \tag{5.5a}$$

$$\text{Prob}(\text{D activated}) = \left|\langle d|\left(\frac{i}{\sqrt{2}}|u\rangle\right)\right|^2 = \frac{1}{4} \tag{5.5b}$$

$$\text{Prob}(\text{photon absorbed by obstruction}) = \left|\langle v|\left(\frac{1}{\sqrt{2}}|v\rangle\right)\right|^2 = \frac{1}{2} \tag{5.5c}$$

If the photon is detected at D, then we know the bomb is active even though it has not been triggered.

5.2.2 A quantum "bomb"

Since the blocking object influences the ultimate nature of the photon detection even though the photon is not detected (absorbed) there, it can be thought of as a "silent detector." Hardy provided a twist (Hardy, 1992b) on the original Elitzur–Vaidman IFM. In his version, the bomb or other macroscopic "silent detector" is replaced by a quantum system: a spin one-half atom. The atom is prepared in a state of spin "up along *x*," which is then subject to a magnetic field gradient along the *z* direction and spatially separated so that it could be found in either of two boxes, one of which ("spin up along *z*," denoted by the state $|z\uparrow\rangle$) is carefully placed in one path of the MZI. (Refer to Figure 5.5.)

As noted in Hardy's discussion and by Elitzur *et al.* (2002), the surprising feature of this experiment is that when detector D is activated, the atom must always be found in the box intersecting path *v*, in a well-defined spin state $|z\uparrow\rangle$; yet seemingly the photon did not interact with it, since the latter was detected at D and therefore was not absorbed by the atom. How is it possible for a photon which apparently went "nowhere near" an atom to dictate the state of the atom? Hardy's discussion is based on the idea of "empty waves," i.e., Bohmian guiding waves in which the Bohmian particle is clearly absent yet the wave appears to have real effects. It is not our purpose here to address the Bohmian "empty wave" picture but to show that TI gives a natural and revealing account of this experiment.

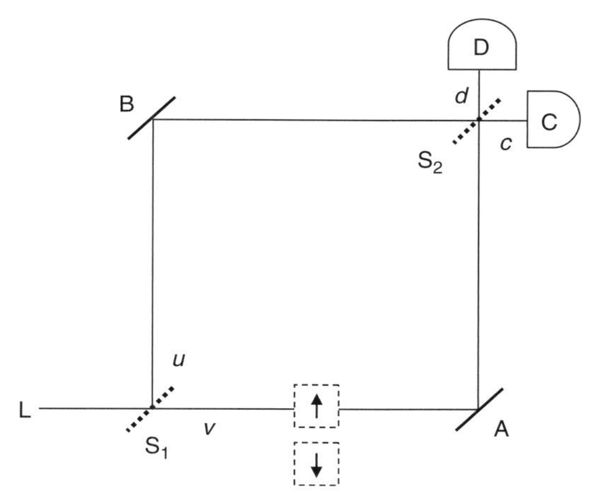

Figure 5.5 Hardy's version of the Elitzur–Vaidman interaction-free measurement with an atom replacing the bomb. *L* denotes a coherent (laser) photon source.

The atom is understood to be in its ground state $|0\rangle$ unless otherwise specified. The atom's excited state – that is, its state when it has absorbed a photon – is denoted as $|1\rangle$. The state of the combined system of {photon, atom} starts out as:

$$|\Psi\rangle_{\mathrm{i}} = |s\rangle \otimes \frac{1}{\sqrt{2}}(|z\uparrow\rangle + |z\downarrow\rangle) \tag{5.5}$$

where $|s\rangle$ denotes the photon source state. As before, the photon's state undergoes a phase shift of *i* upon reflection, so after passing through the first beam splitter S_1, the photon's state becomes $\frac{1}{\sqrt{2}}(i|u\rangle + |v\rangle)$. At this point the total system's state is:

$$\begin{aligned} |\Psi\rangle_{\mathrm{S}_1} &= \frac{1}{2}(i|u\rangle + |v\rangle) \otimes (|z\uparrow\rangle + |z\downarrow\rangle) \\ &= \frac{1}{2}(i|u\rangle|z\uparrow\rangle + |v\rangle|z\uparrow\rangle + i|u\rangle|z\downarrow\rangle + |v\rangle|z\downarrow\rangle) \end{aligned} \tag{5.6}$$

Now, under TI this state represents an offer wave. The second term on the right-hand side of (5.6) involves a potential transaction corresponding to the photon being found on path *v* and the atom occupying the intersecting box. Under the idealized assumptions of the experiment, the atom in the state $|z\uparrow\rangle$ constitutes an absorber for the photon.[18] Note that this experiment presents a further challenge for the 1986 version of TI, in that one of the absorbers is itself in a superposition. Under these circumstances, actualization of the associated

[18] As discussed in Chapter 6, microscopic currents such as atoms have only an amplitude to generate confirmations.

transaction results in actualization not only of the property corresponding to the OW component absorbed, but also of the property corresponding to the state of the absorber involved. Thus, the actualization of this transaction will result in absorption of the photon by the atom, changing it from its unexcited state $|z\uparrow;0\rangle$ to its excited state $|z\uparrow;1\rangle$ and actualizing the photon in state $|v\rangle$.[19] The fact that absorbers can be in superpositions is consistent with PTI's ontology that spacetime grows from outside itself (from "pre-spacetime" – the realm of superpositions, as discussed in Chapter 4), and that the future is non-actualized (as discussed in Chapter 8).[20]

Note that there is a component of the photon OW interacting with the atomic state $|z\uparrow\rangle$ and also a component of the photon OW interacting with the absorbers C and D, since the photon OW component $|v\rangle$ may not be absorbed by the atom (this possibility corresponds to the atomic component $|z\downarrow\rangle$), and may continue on to C and D. Thus the existence of the superposed atomic OW creates a new possibility for the photon: the photon OW component $|v\rangle$ performs "double duty." This underscores the inappropriateness of trying to picture the photon OW as literally propagating "in spacetime." In the latter approach, one expects to follow a photon "trajectory" through arm v; "first" it may encounter an atom in the state $|z\uparrow\rangle$, but it might not, and if it doesn't, the *same* OW continues on to S_2 where it splits into components heading for C and D. But under TI there are CW from the atomic state $|z\uparrow\rangle$ *and* from C and D; these are not mutually exclusive possibilities as implied by the spacetime trajectory story. So we must allow for the photon OW component $|v\rangle$ to be "doing two possible things at once": interacting with the atom in state $|z\uparrow\rangle$ *and* continuing past the atom in state $|z\downarrow\rangle$ to C and D. This illustrates the futility of clinging to a spacetime ontology for OW and CW, which are objects "too big" to fit into spacetime, as argued in Chapter 4. The interactions of OW and CW do not take place on spacetime trajectories. While they are of course constrained by aspects of the experimental arrangement, those constraints take the form of specific, highly probable transactions between the offer waves and the experimental apparatus (such as reflection from a mirror).[21]

[19] Note that the collapse, not only of an OW to one particular component but also of microscopic absorbers in superpositions, can be seen as the way in which events can be actualized in a true "becoming" picture of spacetime. That is, the future absorber is not "already there"; absorbers themselves are only possibilities. Macroscopic absorbers are simply far more probable than microscopic ones such as the atom in the Hardy variation on the Maudlin experiment. I return to the issue of spacetime "becoming" in Chapter 8.

[20] Does this mean that emitters can be in superpositions as well? No, at least in the following sense: when we write down the quantum state for a system, which is an offer wave, that offer wave is always clearly defined and therefore the state of its emitter is well-defined. Any emitter whose state is not well-defined will be epistemically inaccessible to us. However, at the relativistic level, an emitter has only an amplitude to emit, so it is in a kind of "superposition" with respect to a given emission event.

[21] At the relativistic level, reflection is a type of scattering. (See also Feynman, 1985, pp. 101–5.)

If the absorption transaction with the atom does not occur, the second term on the right-hand side of (5.6) is "out of the running" and the system is represented by the state

$$|\Psi\rangle_f = -\frac{1}{2\sqrt{2}}|d\rangle|z\uparrow\rangle + \frac{i}{2\sqrt{2}}|c\rangle|z\uparrow\rangle + \frac{i}{\sqrt{2}}|c\rangle|z\downarrow\rangle \tag{5.7}$$

(In (5.7), the overall factor of $1/\sqrt{2}$ arises from the atomic state.) The terms involving detection at C involve ambiguous states of the atom; I disregard these and focus our attention on the interesting case which is detection at D, represented by the first term in (5.7). As the photon component of this offer wave is absorbed by D, a photon CW is produced of the same initial amplitude, $\frac{1}{2}\langle d|$. This photon CW component interacts with the two beam splitters, thus acquiring another factor of 1/2 along the way before encountering the photon source *S*, for a final photon CW amplitude of 1/4. Meanwhile, the atomic CW, $\frac{1}{\sqrt{2}}\langle z\uparrow|$, picks up another factor of $1/\sqrt{2}$ due to being split by the Stern–Gerlach field which transformed the original OW from the state $|x\uparrow\rangle$ into equal components of "up along *z*" and "down along *z*," for a final atomic CW amplitude of 1/2. (Only the component of the CW that matches the atomic source state $|x\uparrow\rangle$ can participate in the possible transactions for the system.)

Thus the total CW amplitude at the photon and atom emitters is $\left(\frac{1}{4}\right)\left(\frac{1}{2}\right) = \left(\frac{1}{8}\right)$, in agreement with standard predictions. The form of the combined OW (5.7) ensures that a D transaction can only occur for an atom in the "silent detector" state $\langle z\uparrow;0|$, which explains why the atom's initial superposition must be "collapsed" whenever the photon is detected at D.

5.3 The Hardy experiment II

The "Hardy experiment II" (Hardy, 1992a), not to be confused with Hardy's variation on the basic Elitzur–Vaidman IFM described above, presents an apparent paradox based on a combination of *two* IFMs. Recall that in the basic IFM, a Mach–Zehnder interferometer (MZI) is tuned so that one of the two detectors, a "silent detector" typically labeled D, will never be activated unless there is an obstruction in one of the arms (referred to in what follows as the "blocking" arm and labeled *v*; see Figure 5.6). This experiment uses two overlapping MZIs, one for an electron and the other for a positron. Hardy's idealized presentation assumes that if the electron and positron meet in the overlapping region corresponding to both blocking arms, they will annihilate with certainty. (This is not strictly speaking correct, of course, since there is an amplitude less than unity for electron-position annihilation into two photons to occur no matter how close the

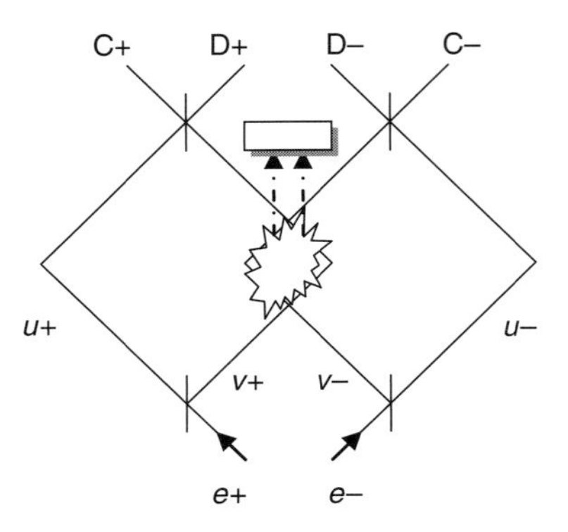

Figure 5.6 A schematic diagram of the Hardy experiment.

particles get.[22]) It turns out that even in cases where there is no annihilation (i.e., both particles are "not in" the overlapping arms corresponding to the term $|v+,v-\rangle\rangle$, both detectors D+ and D– can activate. This outcome theoretically occurs with a probability of 1/16. If we think of quanta as having definite whereabouts in the apparatus at all times, this seems paradoxical, since each individual apparatus is supposedly only able to have its D detector activated when something is blocking arm v. The event of both detectors D activating therefore seems to imply that both quanta must be in arms v+ and v–, but then they should annihilate (or at least be mutually scattered), so presumably could not reach the detector area at all. Hence the paradox.

However, the above is only a paradox if we insist on thinking of quantum objects as classical corpuscles carrying energy and momentum along specific trajectories. This classical "billiard ball" story mistakenly tells us that an amplitude for an interaction to occur somewhere (e.g., in the blocking arm of the MZI) means that a corpuscle must actually be physically present there if some other detection (e.g., at D) occurs which depends on the given amplitude. This "billiard ball" notion is what is denied in TI: quanta are not corpuscles pursuing trajectories. Amplitudes describe offer and confirmation waves which themselves do not transfer energy, but which can give rise to transactions. It is the completed (actualized) transactions that transfer energy and other conserved quantities, and which can therefore activate detectors.

5.3.1 Details of the Hardy experiment II

Consider Figure 5.6, which illustrates the setup for Hardy's experiment. The state of a quantum after passing the first beam splitter (a half-silvered mirror indicated in the figure by a short vertical line) is[23]

[22] Cf. Berestetskii *et al.* (2004), pp. 368–70.

[23] Reflections result in a phase change of $\pi/2$, or a factor of i, for the component reflected, as discussed in Section 5.2.1.

$$|\psi_1\rangle = \frac{1}{\sqrt{2}}[|u\rangle + i|v\rangle] \tag{5.8}$$

Subsequently, each of the components v and w evolves as follows through the second beam splitter (the labels c and d refer to paths leading to the respective detectors C and D):

$$|u\rangle \to \frac{1}{\sqrt{2}}[|d\rangle + i|c\rangle] \tag{5.9a}$$

$$|v\rangle \to \frac{1}{\sqrt{2}}[|c\rangle + i|d\rangle] \tag{5.9b}$$

So that the state $|\psi_1\rangle$ evolves to $i|c\rangle$ when there is nothing obstructing either arm of the MZI.

The total system's state just after the first beam splitter is:

$$|\Psi_1\rangle = \frac{1}{2}[|u+\rangle + i|v+\rangle] \otimes [|u-\rangle + i|v-\rangle]$$

$$= \frac{1}{2}[|u+,u-\rangle + i|u+,v-\rangle + i|v+,u-\rangle - |v+,v-\rangle] \tag{5.10}$$

where the kets with two labels are elements of the 4-dimensional Hilbert space of the combined system.

The fourth term in (5.10) represents electron–positron annihilation in Hardy's idealization, which assumes that e+ and e– annihilate with certainty into two photons when they are both in the overlapping region. The two photons, indicated by the upward dotted arrows, are absorbed by a detector (indicated by the shadowed rectangle). If the two quanta were non-annihilating objects, such as two (coherent[24]) photons, the total state would simply evolve to $-|c+,c-\rangle$ and all quanta would be detected at C+,–. However, with the fourth term absent (i.e., according to the idealization, in cases where the electron and positron do not annihilate), we need to follow the evolution of the remaining three terms to see what detections are possible. Considering only the amplitudes of the component $|d+,d-\rangle$ for times after the

[24] The two quanta have to be perfectly in phase for cancellation to occur.

second beam splitter, we find the following contributions from the first three terms in (5.10):

$$|u+,u-\rangle \quad \text{gives} \quad \frac{1}{4}|d+,d-\rangle \tag{5.11a}$$

$$|u+,v-\rangle \quad \text{gives} \quad -\frac{1}{4}|d+,d-\rangle \tag{5.11b}$$

$$|v+,u-\rangle \quad \text{gives} \quad -\frac{1}{4}|d+,d-\rangle \tag{5.11c}$$

(and note that, if annihilation were not possible, the fourth term would give the same contribution as $|u+,u-\rangle$, thus canceling all contributions of $|d+,d-\rangle$).

Thus the fact that the three remaining terms contribute a non-zero amplitude for $|d+,d-\rangle$ (specifically, an amplitude of ¼) makes the detection at D+,D– possible when discounting contributions from the term $|v+,v-\rangle$, which leads to annihilation. Now let us see how TI describes this experiment.

5.3.2 The TI account

First, recall that according to TI, transfers of energy resulting in detection occur only as a result of actualized transactions (as reviewed in Chapters 3 and 4). Yet there is much that goes on "behind the scenes" leading up to a transaction.[25] The following are necessary (but not sufficient) conditions: first, an offer wave is emitted. In the Hardy experiment there are two single-quantum OWs corresponding to the electron and positron. The OWs propagate until they encounter a possible absorber. Thus the first opportunity for absorption corresponds to the term $|v+,v-\rangle$, in which the two OWs may encounter each other. As mentioned earlier, an accurate treatment of this situation would consider the relativistic scattering cross-section for e+,e– annihilation, but let's restrict the discussion to the non-relativistic idealization presented by Hardy and assume that this term is equivalent to annihilation of the e+,e– OW and the generation of two photon OWs. In this case, an *incipient* (possible) transaction is established in virtue of confirmation waves generated by the detector for the photons. The generation of confirmation waves is a necessary

[25] The apparently "pseudo-temporal" language here refers roughly analogous to Cramer's "pseudotime," except that that was a heuristic term whereas the possibility space discussed here is considered to be a genuine physical realm beyond spacetime.

condition for a transaction, but as noted above, not sufficient. The "choice" of which transaction is realized is irreducibly stochastic (i.e., there is no determinate sufficient condition for a transaction). The probability of the annihilation transaction is given by the product of the OW and CW amplitudes, or ¼.

If this annihilation transaction does not occur, there is still an OW component for the combined system corresponding to a $|d+,d-\rangle$ transaction, with an amplitude of ¼ (see (5.11)). That is, OWs for each quantum (e+ and e–) can reach detectors D+,–. The detectors are composed of absorbers which respond to each OW by returning a CW of the same amplitude to the respective emitters (e+ and e–). The probability of this transaction is the final amplitude of this CW, which undergoes the same attenuation (through interactions with components of the apparatus such as beam splitters; see Cramer, 1986, pp. 661–2, 674–5) as the original OW; the final amplitude is given by the Born Rule, (¼) (¼) = 1/16.

Thus there is nothing paradoxical if we see these processes as involving interactions between offer and confirmation waves rather than as dictating the supposed whereabouts of localized particles. However, this experiment, like the previous one, involves one or more OW components having to do "double duty." In the Hardy experiment II, this situation occurs because the single-quantum components corresponding to the "blocking" arm, $|v+\rangle$ and $|v-\rangle$, are still needed for transactions involving detectors C and D. The latter possibilities arise from the terms (5.11b) and (5.11c).

So we can't just say that if the annihilation transaction corresponding to the term $|v+,v-\rangle$ doesn't occur, then the entire content of that term is "out of the picture," because we still need its single-particle components. If we try to hold onto a picture of single-particle waves propagating through the apparatus, we end up with an awkward account in which, e.g., the positron OW component $|v+\rangle$ "decides" not to engage in a transaction placing it in the blocking arm (corresponding to $|v+,v-\rangle$), but still has to be "present" in the blocking arm (corresponding to $|v+,u-\rangle$) to give the correct contribution to detection at D–, as does the electron OW component corresponding to $|u+,v-\rangle$. Both these terms must remain in play, and both imply that the electron and positron OW have some presence "in the overlapping arm," but this contradicts the premise that annihilation must occur in that case; for annihilation is already out of the picture at this point in the analysis. So these states can't consistently be taken as describing OW components actually present within spacetime in one arm or the other.

The resolution of this puzzle is the same as discussed in the previous section: we cannot picture the entities described by quantum states as literally propagating *in* spacetime through the arms of an MZI. Instead, quantum states describe dynamical *possibilities* whose domain is mathematically described by Hilbert space, not 3-space or spacetime. Spacetime is the theater of completed transactions, not the domain of quantum states which requires a larger mathematical structure (because there are enormously more possibilities than can be actualized in spacetime). An accurate description of an experiment involving microscopic systems which require a quantum mechanical description must treat the *entire* experimental apparatus and quantum system as a nexus of OW, CW, incipient transactions, and actualized transactions. The macroscopic features of the apparatus will correspond to highly probable and persistent transactions which enable it to be thought of as "existing in spacetime," since spacetime is the domain of the structured set of actualized (successful) transactions. However, elements of the experiment with significantly fewer and less probable transactions (the electron and positron in this case) do not really exist in spacetime but interact with the relevant aspects of the apparatus (i.e., absorbers) on the level of possibility (OW and CW), in *the larger possibility space* corresponding to all the quanta comprising the entire system.

5.4 Quantum eraser experiments

The term "quantum eraser experiment" refers to a class of experiments involving a pair of correlated photons. One of the pair is termed the "signal" photon and the other is termed the "idler" photon. The signal photons are directed into a two-slit apparatus and, depending on what is done with their paired idler photons, an interference pattern may or may not be seen for the signal photons. (Some versions of the experiment send a single photon through the two-slit apparatus and then convert it into two correlated photons after the two slits; this is the version discussed below.) "Erasing" refers to the process in which a particular kind of measurement of the idler photon obliterates the so-called "which-slit" information associated with the signal photon. There are separate detection arrangements for the signal and idler photons, and their separate detection information is sent to a coincidence counter to keep track of the pairs.

5.4.1 Details of a quantum eraser experiment

The signal photons in this type of experiment are always detected at a detector S which is scanned across positions x to determine the count at each position (refer to Figure 5.7). That information is sent to the coincidence counter. The idler photons may be subjected to (1) a "which-slit" measurement or (2) a "both-slits" measurement, with two detectors corresponding to each of (1) and (2). Idler detections are sent to the coincidence counter which provides the information about which signal detections are paired with which idler detections.

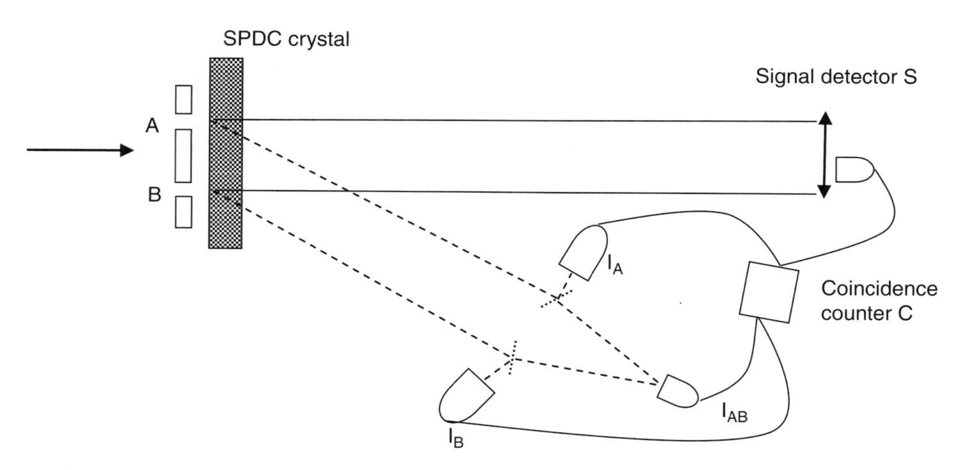

Figure 5.7 A quantum eraser experiment. [I am indebted to Ross Rhodes for suggestions for this and the following figure.]

Any idler photon that activates detectors (1) is correlated with a signal photon with "which-slit" information, and any idler photon that activates detectors (2) reflects a superposition of slit states. It is only by looking at appropriate coincidences that one can see the above effects.

Recall from Chapter 3 that one can do a two-slit experiment with one type of particle (in that case, an electron) along with an auxiliary measurement by another type of particle (in that case, a photon). In that example, the electron played the part of the "signal photon" and the photon played the part of the "idler." What the experimenter chooses to do with the photon (i.e., how sharp a measurement to make) affects whether or not electron self-interference takes place (i.e., whether or not one sees an interference pattern for the electrons or a distribution corresponding to definite slit paths). Quantum eraser experiments extend that basic setup by replacing the choice of how sharp a measurement to make with a choice of what kind of process is imposed on the idler.

In the usual approach to discussing these types of experiments, it is assumed that the signal photon either "went through a particular slit" or "went through both slits," depending on the kind of measurement performed on the idler photon. This seems to imply the very mysterious idea that what is done with the idler photon can materially affect the signal photon's spacetime trajectory. However, in TI, the influences involved are not at the level of "we poked one photon and somehow ended up instantly (or even retroactively!) poking another photon in a completely different part of the experimental apparatus." This is because in TI the photon is not a corpuscle pursuing a spacetime trajectory. Rather, the OW is a physical possibility created by the source together with the two-slit configuration, and that OW has a particular state – in this case, the two-slit (two-photon) state, irrespective of what kind of measurement is made. So in all these variations on the two-slit experiment, the OW is a "both-slit" entity. The possible transactions available to that entity depend on the absorber configuration which generates CWs.[26]

The experimental setup of the version by Kim *et al.* (2000) is depicted schematically in Figure 5.7. The original OW, which can be written as

$$|\psi\rangle = \left(1/\sqrt{2}\right)[|A\rangle + |B\rangle] \tag{5.12}$$

is converted into a two-photon correlated OW by way of a "spontaneous parametric down conversion" (SPDC) process. This process duplicates each "which-slit" component but with opposite polarizations for each of the two photons. If we don't explicitly write the polarization states (which serve to correlate the two photons and enable experimenters to send them off into different directions), the two-photon state can be written as

$$|\Psi\rangle = \left(1/\sqrt{2}\right)[|A\rangle|A\rangle + |B\rangle|B\rangle] \tag{5.13}$$

where the first and second kets in each term correspond to the signal and idler, respectively. The signal photon OW components are sent to detector S and the idler photon OW components to the detector assembly, which is actually a system of beam splitters and mirrors with four subdetectors I_A, I_B, I_{AB} and I_{AB}'. (I_{AB}' is not shown in the diagram for simplicity.) The latter two both detect an interference pattern; they are just shifted by a phase of π with respect to each other. (The sum of the patterns for I_{AB} and I_{AB}' is the same as the sum of the patterns for I_A and I_B.)

Figure 5.7 schematically shows the idler detectors I_A, I_B, I_{AB}. The beams corresponding to passage through A and B are split by half-silvered mirrors. The reflected component of each is sent to detectors I_A and I_B, respectively, which provides a "which-path" measurement of its signal photon partner (just as in the use of telescopes aimed at each slit in the two-slit experiment); and the transmitted components of each will be recombined and may reach the other two detectors I_{AB} and I_{AB}'. The recombined A and B beam components detected by I_{AB} and I_{AB}' have no "which-path" information, and this is the "quantum eraser."

[26] The phrase "absorber configuration" here includes all components of the experiment such as reflecting components. Such interactions need to be described at the relativistic level for accuracy, and involve scattering. The relativistic aspect of PTI is discussed in Chapter 6.

Meanwhile, the signal photon heads toward the movable detector S, which is located at varying positions x for different runs of the experiment. If the signal photon is detected at position x, detector S sends a count to the coincidence counter. The idler detection for that run, wherever it occurs, is matched via the coincidence counter to its partner signal photon. (If the signal photon is not detected at x, it cannot be matched to its idler partner and that run does not show up in the coincidence count.) In this way, the experimenters have a joint count: that is, for all signal photon detections at position x, they can see how many idler photons were detected at each of the idler detectors. Those signal photons whose idlers were detected at I_A and I_B turn out (as predicted by standard quantum mechanical calculations of the relevant probabilities) to be distributed in a non-interfering "single-slit" distribution, while those whose idlers were detected at I_{AB} form an interference pattern – the "which-slit" information has been "erased."

5.4.2 The TI account

The TI account of this experiment is as follows: the total system's OW is as given in (5.13). Detector S generates the CW $\langle x|$ corresponding to its position in any given run, and the signal photon may therefore be absorbed at S(x), where this notation specifies the position of S for any given run. However, the signal photon does not have a well-defined OW since it is a single component of an entangled two-photon state; the OW is well-defined only for the two-photon state. In fact, this is the familiar fact that the state of a component system of a larger, multi-component system is an "improper mixture." The component can be represented not by a pure state but only by a mixed state or "density operator"; this will be discussed further below. Meanwhile, it is the existence of different possible CW for the idler photon, based on the splitting of the idler component of the original OW into either a "which-slit" or "both-slits" detector region, that makes possible different sets of incipient transactions for the idler: either "which-slit" transactions or "both-slit" transactions.

The experiment may also be implemented with a delayed aspect (see Figure 5.8): the idler photon measurement may be delayed until after the signal photon has been detected at S. This makes the experiment seem astounding from the standard point of view. A typical discussion of a variation of the delayed version says, in part:

Before photon p [the idler photon] can encounter the [erasing] polarizer, s [the signal photon] will be detected. Yet it is found that the interference pattern is still restored. It seems s knows the "which-way" marker has been erased and that the interference behavior should be present again, without a secret signal from p. How is this happening? It wouldn't make sense that p could know about the polarizer before it got there. It can't "sense" the polarizer's presence far away from it, and send photon s a secret signal to let s

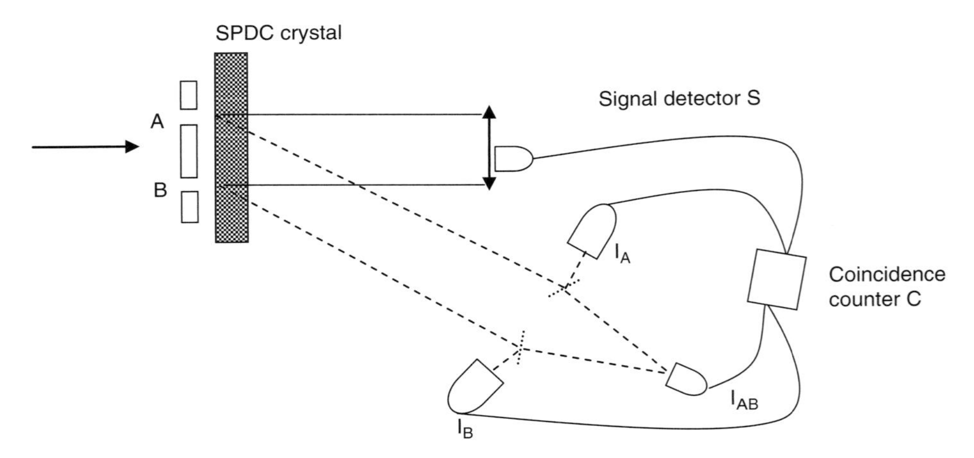

Figure 5.8 Delayed quantum erasure.

know about it. Or can it? And if photon p is sensing things from far away, we shouldn't assume that photon s isn't."[27]

The above discussion includes the usual metaphysical assumption that I believe needs to be rejected; i.e., that an emitted photon is pursuing a spacetime trajectory. This makes the phenomena seem particularly bizarre, necessarily involving remote sensing and/or foreknowledge on the part of photons considered as material corpuscles. Meanwhile, the TI account of the delayed choice version of this experiment is not fundamentally different from its standard version and simply involves a set of transactional opportunities. First, the signal photon may be detected at S(x) with a probability of $P(x) = Tr_S(|x\rangle\langle x|\rho_S)$, where ρ_S is the "reduced" density operator for the signal photon, $\rho_S = Tr_I(\Psi\rangle\langle\Psi| = ½[|A_S\rangle\langle A_S| + |B_S\rangle\langle B_S|]$. (I augment the kets with a label S or I for signal or idler respectively, for clarity.) The notation Tr_S or Tr_I means that the trace is a partial one taken over the corresponding component of the composite state.[28] The signal photon absorption at x constitutes the actualization of a component of an incipient two-particle

[27] Excerpted from Orozco (2002).
[28] As discussed in Chapter 3, the trace is just a way of saying "We know that, given a set of absorbers corresponding to a particular basis, a set of transactions are possible, and those are not permitted between non-matching OW and CW." The weights of all possible transactions for a given basis always sum to unity. The basis-independence of the trace reflects its status as a "wild card" or placeholder for unknown or unspecified absorber configurations. The novel feature of the quantum eraser is that the idler detection basis itself is uncertain based on the splitting of the idler OW as it encounters the beam splitter; the four OW components reaching the detectors I are thus attenuated by a factor of (1/√2), and therefore so are the CW from all detectors I. So in this case, the weights of all four idler transactions (given a particular signal detection at S(x)) sum to unity as well.

transaction corresponding to a particular *attenuated* two particle OW; let's call that OW component $|\Psi_{\mathrm{ATT}}\rangle$:

$$|\Psi_{\mathrm{ATT}}\rangle = (1/\sqrt{2})[\langle x|\mathrm{A}_S\rangle|x\rangle|\mathrm{A}_I\rangle + \langle x|\mathrm{B}_S\rangle|x\rangle|\mathrm{B}_I\rangle] \tag{5.14}$$

This is analogous to the attenuation of a single-particle OW by a polarizing filter. In that case, if the particle is not absorbed by the filter, only the OW component passing the filter continues on to a final detection. In the two-particle situation here, one of the particles (the signal photon) is detected at S_x and its companion particle's OW (now a pure state) continues on to a final detection. Note that this corresponds to eliminating the "improper mixed state" for both particles; once either one of them has been absorbed by way of a one-particle transaction, the other particle is now in a pure state. This is evident in the fact that we can factor out the common factor $|x\rangle$, rewriting (5.14) as a product state

$$|\Psi_{\mathrm{ATT}}\rangle = (1/\sqrt{2})|x\rangle[\langle x|\mathrm{A}_S\rangle|\mathrm{A}_I\rangle + \langle x|\mathrm{B}_S\rangle|\mathrm{B}_I\rangle] \tag{5.15}$$

so the attenuated idler OW is just

$$|\Psi^I_{\mathrm{ATT}}\rangle = (1/\sqrt{2})[\langle \mathrm{x}|\mathrm{A}_S\rangle|\mathrm{A}_I\rangle + \langle \mathrm{x}|\mathrm{B}_S\rangle|\mathrm{B}_I\rangle] \tag{5.16}$$

where the inner products are just complex numbers a and b. We must keep in mind that $|\Psi^I_{\mathrm{ATT}}\rangle$ is further split, via the beam splitters, into distinct components going to detectors $\mathrm{I_A}$, $\mathrm{I_B}$ and to $\mathrm{I_{AB}}$, $\mathrm{I_{AB}}'$. Calculations of probabilities for detections at each set of detectors are therefore based on this further attenuation of $|\Psi^I_{\mathrm{ATT}}\rangle$.

After the signal photon is absorbed, there are still one-particle transactional opportunities available for the idler photon which is now in the pure state (5.16); but the attenuation of the original two-particle OW makes the weights of the transactional opportunities available to the idler photon conditional on the attenuated OW component (5.16), rather than on the original unmodified $|\Psi\rangle$.

As in the Wheeler delayed choice experiment discussed previously, the "which-slit" or "both-slits" aspect of the idler CW is indeterminate, and it is this indeterminacy that prevents the observation of a clear pattern at detector S without the choice of a well-defined subset of runs from the coincidence counter. The choice of which type of idler detections ("which slit" or "both slits") to select from the coincidence counter selects the ensemble of signal photon transactions corresponding to a particular type of idler CW, regardless of whether the signal photon is detected before or after the idler measurement is selected.

In the non-delayed version, the absorption of an idler photon at one of the "which-slit" detectors makes the weight of the signal photon's incipient transaction for absorption at S(x) conditional on the prior transaction with the idler. For example, if the idler photon is detected at I_A, this means the transaction corresponding to the two-particle component $|A\rangle|A\rangle$ has been actualized. Therefore, the weight of the signal photon incipient transaction for absorption at S(x) is much higher for x corresponding to the peak of the "A-slit" distribution than it is for x' corresponding to, say, the peak of the "B-slit" distribution, since the two-particle transaction for the latter state has failed.

Some further remarks are in order regarding the issue of an improper mixture. In a composite system described by a pure state (such as (5.13)), the improper mixed state of a component system can be considered analogous to the basis arbitrariness of a pure state for a single system whose absorption opportunities have not been specified. One may write a single-system state in terms of any basis, but that leads to the measurement problem – i.e., it provides no way to say how, when, or why a "measurement" corresponding to a particular basis has occurred. (This is so even if one takes into account a unitary interaction correlating the system with a "measurement apparatus," since that combined state can also be written in any basis.) This problem is often considered partially remedied by Bohr's "Copenhagen interpretation," which specifies that the "entire experimental arrangement" must be taken into account, but the weakness of the CI is that there is no physical reason for this, and inevitably a "conscious observer" has to be brought in, with the ensuing psycho-physical speculations. As noted in Chapter 1, decoherence approaches attempt to remove basis ambiguity by specifying a "cut" between the system and its environment, but that also is dependent on a stipulation of the distinction between the environment and the system that presupposes the perceptions and intentions of an outside observer. TI provides the physical reason for the necessity of taking into account the experimental arrangement: absorption is a physical process, and that is what takes place in virtue of the experimental arrangement.

Returning to the improper mixed state, the "impropriety" of the mixed state of a component system reflects the fact that it actually does not possess its own OW component. Nevertheless, *individual CW* may be present for the component systems, as is the case with the quantum eraser and in the more familiar EPR–Bohm experiment (see Appendix C for a more detailed discussion). This makes possible a set of single-particle incipient transactions, the existence of which then dictates the physical basis applying to the entire system. The basic point is that *proper* mixtures and well-defined bases for systems are always defined with reference to an absorber basis. (Of course, the absorber basis itself may not be well-defined, as in the quantum eraser experiment, but a transactional account can still be given as above; when a transaction is actualized corresponding to absorption of a component system, the other system's OW becomes well-defined with respect to that basis.)

This concludes my study of specific experimental challenges to TI and applications of PTI. In the remaining chapters, I explore the relativistic domain and consider further metaphysical implications of the interpretation.

6

PTI and relativity[1]

6.1 TI and PTI have basic compatibility with relativity

As noted in Cramer (1986), the original version of TI already has basic compatibility with relativity in virtue of the fact that the realization of a transaction occurs with respect to the endpoints of a spacetime interval or intervals, rather than at a particular instant of time, the latter being a non-covariant notion. Its compatibility with relativity is also evident in that it makes use of both the positive and negative energy solutions obtained from the Schrödinger equation and the complex conjugate Schrödinger equation respectively, both of which are obtained from the relativistic Klein–Gordon equation by alternative limiting procedures. Cramer (1980, 1986) has noted that, in addition to Wheeler and Feynman, several authors (including Dirac) have laid groundwork for and/or explored explicitly time-symmetric formulations of relativistic quantum theory with far more success than has generally been appreciated.[2] This chapter is largely devoted to developing PTI in terms of a quantum relativistic extension of the Wheeler–Feynman theory by Davies (1970, 1971, 1972). First, I present some preliminary remarks.

6.1.1 Emission and absorption are fundamentally relativistic processes

The crucial feature of TI/PTI that allows it to "cut the Gordian knot" of the measurement problem is that it interprets absorption as a real physical process that must be included in the theoretical formalism in order to account for any measurement result (or more generally, any determinate outcome associated with a physical system or systems). The preceding is a specifically relativistic aspect of quantum theory, since non-relativistic quantum mechanics ignores absorption: it

[1] Much of the material in this chapter is based on a paper forthcoming in *Foundations of Physics*, entitled "The possibilist transactional interpretation and relativity."

[2] For example, Dirac (1938), Hoyle and Narlikar (1969), Konopinski (1980), Pegg (1975), Bennett (1987).

addresses only persistent particles. Strictly speaking, it ignores emission as well; there is no formal component of the non-relativistic theory corresponding to an emission process. The theory is applied only to an entity or entities assumed to be already in existence. In contrast, relativistic quantum field theory explicitly includes emission and absorption through the field creation and annihilation operators respectively; there are no such operators in non-relativistic quantum mechanics.[3] Because the latter treats only pre-existing particles, the actual emission event is not included in the theory, which simply applies the ket $|\Psi\rangle$ to the pre-existing system under consideration. Under these restricted circumstances, it is hard to see a physical referent to the brac $\langle\Psi|$ from within the theory, even though it enters computations needed to establish empirical correspondence. What TI does is to "widen the scope" of non-relativistic quantum theory to take into account both emission and absorption events, the latter giving rise to the advanced state or brac $\langle\Psi|$. In this respect, again, it is harmonious with relativistic quantum theory.

6.1.2 TI/PTI retains isotropy of emission (and absorption)

It should also be noted that the standard notion of emission as being isotropic with respect to space (i.e., a spherical wave front) but *not* isotropic with respect to time (i.e., that emission is only in the *forward* light cone) seems inconsistent, and intrinsically ill-suited to a relativistic picture, in which space and time enter on an equal footing (except, of course, for the metrical sign difference). The prescription of the time-symmetric theory for half the emission in the $+t$ direction and half in the $-t$ direction is consistent with the known fact that emission does not favor one spatial direction over another, and harmonious with the relativistic principle that a spacetime point is a unified concept represented by the four-vector $x^\mu = \{x^0, x^1, x^2, x^3\}$. This symmetry principle, and the consistency concern related to it, rather than a desire to eliminate the field itself (historically the motivation for absorber-based electrodynamics, see below), is the primary motivation for TI in its relativistic application.

6.2 The Davies theory

We turn now to the theory of Davies, which provides a natural framework for PTI in the relativistic domain.

[3] Technically, the Davies theory, which is probably the best currently articulated model for TI and which is discussed below, is a direct action (DA) theory in which field creation and destruction operators for photons are superfluous; the electromagnetic field is not really an independent entity. Creation and annihilation of photons is then physically equivalent to couplings between the interacting charged currents themselves, and it is the coupling amplitudes that physically govern the generation of offers and confirmations. The important point is that couplings between fields are inherently stochastic and so are the generations of OW and CW.

6.2.1 Preliminary remarks

The Davies theory has been termed an "action at a distance" theory because it expresses interactions not in terms of a mediating field with independent degrees of freedom, but rather in terms of direct interactions between currents.[4] As Cramer (1986) notes, one of the original motivations for such an "action at a distance" theory was to eliminate troubling divergences, stemming from self-action of the field, from the standard theory; thus it was thought desirable to eliminate the field as an independent concept. However, it was later realized that some form of self-action was needed in order to account for such phenomena as the Lamb shift (although the Davies theory does allow for self-action in that a current can be regarded as acting on itself in the case of indistinguishable currents (see, e.g., Davies (1971), p. 841, figure 2).

Nevertheless, despite its natural affinity for a time-symmetric model of the field, it must be emphasized that PTI does *not* involve an ontological elimination of the field. On the contrary, the field remains at the "offer wave" level. This is the same picture in which the classical Wheeler–Feynman electromagnetic retarded field component acts as a "probe field" that interacts with the absorber and prompts the confirming advanced wave, which acts to build up the emitter's retarded field to full strength and thus enable the exchange of energy between the emitter and the absorber.

Thus PTI is based, not on elimination of quantum fields, but rather on the time-symmetric, transactional character of energy propagation by way of those fields, and the assumption that offer and confirmation waves capable of resulting in empirically detectable transfers of physical quantities only occur in couplings between field currents. However, in keeping with this possibilist reinterpretation, the field operators and field states themselves are considered as pre-spacetime objects. That is, they exist; but not in spacetime. What exist in spacetime are actualized, measurable phenomena such as energy transfers. Such phenomena are always represented by real, rather than complex or imaginary, mathematical objects. At first glance this ontology may seem strange; however, when one recalls that such standard objects of quantum field theory as the vacuum state $|0\rangle$ have no spacetime arguments and are maximally non-local,[5] it seems reasonable to suppose that such objects exist, but not in spacetime (in the sense that they cannot be associated with any region in spacetime).

A further comment is in order regarding PTI's proposal that spacetime is emergent rather than fundamental. In the introductory chapter to their classic *Quantum Electrodynamics*, Beretstetskii, Lifschitz, and Petaevskii make the following observation concerning QED interactions:

[4] The term "current" in this context denotes the generalization of a probability distribution for a particle associated, in the relativistic domain, with a quantum field.

[5] This is demonstrated by the Reeh–Schlieder Theorem; cf. Redhead (1995).

For photons, the ultra-relativistic case always applies, and the expression [$\Delta q \sim \hbar/p$], where Δq is the uncertainty in position, is therefore valid. This means that the coordinates of a photon are meaningful only in cases where the characteristic dimension of the problem is large in comparison with the wavelength. This is just the "classical" limit, corresponding to geometric optics, in which the radiation can be said to be propagated along definite paths or rays. In the quantum case, however, where the wavelength cannot be regarded as small, the concept of coordinates of the photon has no meaning...

The foregoing discussion suggests that the theory will not consider the time dependence of particle interaction processes. It will show that in these processes there are no characteristics precisely definable (even within the usual limitations of quantum mechanics); *the description of such a process as occurring in the course of time is therefore just as unreal as the classical paths are in non-relativistic quantum mechanics*. The only observable quantities are the properties (momenta, polarization) of free particles: the initial particles which come into interaction, and the final properties which result from the process. [The authors then reference L. D. Landau and R. E. Peierls, 1930.[6]] (Emphasis added.) (1971, p. 3)

The italicized sentence asserts that the interactions described by QED (and, by extension, by other interacting field theories) cannot consistently be considered as taking place in spacetime. Yet they do take place *somewhere*; the computational procedures deal with entities implicitly taken as ontologically substantive. This "somewhere" is just the pre-spatiotemporal, pre-empirical realm of possibilities proposed in PTI. The "free particles" referred to in the last sentence of the excerpt exist within spacetime, whereas the virtual (unobservable) particles do not.

6.2.2 Specifics of the Davies theory

The Davies theory (1970, 1971, 1972) is an extension of the Wheeler–Feynman time-symmetric theory of electromagnetism to the quantum domain by way of the S-matrix (scattering matrix). This theory provides a natural framework for PTI in the relativistic domain. The theory follows the basic Wheeler–Feynman method by showing that the field due to a particular emitting current $j^{\mu}_{(i)}(x)$ can be seen as composed of equal parts retarded radiation from the emitting current and advanced radiation from absorbers. Specifically, using an S-matrix formulation, Davies replaces the action operator of standard QED,

$$J = \sum_i \int \mathrm{d}x j^{\mu}_{(i)}(x) A_{\mu}(x) \tag{6.1}$$

[6] The Landau and Peierls paper has been reprinted in Wheeler and Zurek (1983).

(where A_μ is the standard quantized electromagnetic field), with an action derived from a direct current-to-current interaction,[7]

$$J = -\frac{1}{2}\sum_{i,j}\int dx\, dy j^{\mu}_{(i)}(x) D_F(x-y) j_{(j)\mu}(y) \tag{6.2}$$

where $D_F(x-y)$ is the Feynman photon propagator. (This general expression includes both distinguishable and indistinguishable currents.)

While $D_F(x-y)$ implies a kind of asymmetry in that it only allows positive frequencies to propagate into the future, Davies shows that for a "light-tight box" (i.e., no free fields), the Feynman propagator can be replaced by the time-symmetric propagator $\bar{D}(x) = \frac{1}{2}[D^{\mathrm{ret}}(x) + D^{\mathrm{adv}}(x)]$, where the terms in the sum are the retarded and advanced Green's functions (solutions to the inhomogeneous wave equation).

Specifically, Davies shows that if one excludes scattering matrix elements corresponding to transitions between an initial photon vacuum state and final states containing free photons, his time-symmetric theory, based on the time-symmetric action $J = -\frac{1}{2}\sum_{i,j}\int dx\, dy j^{\mu}_{(i)}(x)\bar{D}(x-y) j_{(j)\mu}(y)$, is identical to the standard theory. (See Davies (1972), equations (7)–(10) for a discussion of this point, including the argument that if one considers the entire system to be enclosed in a light-tight box, this condition holds.) The excluded matrix elements are of the form $\langle n|S|0\rangle$, where n is different from zero. By symmetry, for emission and absorption processes involving (theoretically)[8] free photons in either an initial or final state, one must use D_F instead of $\bar{D}$ to obtain equivalence with the standard theory.

To understand this issue, recall Feynman's remark that if you widen your area of study sufficiently, you can consider all photons "virtual" in that they will always be emitted and absorbed somewhere.[9] He illustrated this by an example of a photon propagating from the earth to the moon (see Figure 6.1).

But, as Davies notes, this picture tacitly assumes that real (not virtual) photons are available to provide for unambiguous propagation of energy *from* the earth *to* the moon. If such free photons are involved, then (at least at the level of the system in the drawing) we don't really have the light-tight box condition allowing for the use of $\bar{D}$ rather than D_F. (In any case, $\bar{D}$ alone would not provide for the propagation of energy

[7] That these expressions are equivalent is proved in Davies (1971) and reviewed in (1972). The currents j^μ are fermionic currents, such as $\bar{u}\gamma^\mu u$.

[8] The caveat "theoretically" is introduced because a genuinely free photon can never be observed: any detected photon has a finite lifetime (unless there are "primal" photons which were never emitted) and is therefore not "free" in a rigorous sense. This is elaborated below and in note 10.

[9] Feynman (1998). Sakurai (1973, p. 256) also makes this point.

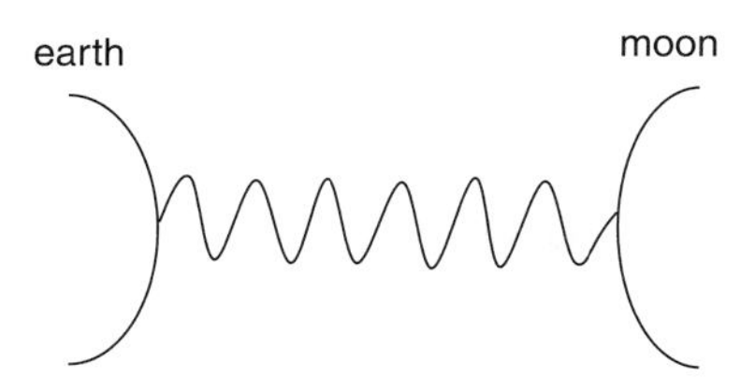

Figure 6.1 A "virtual" photon propagating from the earth to the moon.

in only one direction; time-symmetric energy propagation in a light-tight box in an equilibrium state would be fully reversible. Thus the observed time-asymmetry of radiation must always be explained by reference to boundary conditions, either natural or experimental.) So one cannot assume that equivalence with the standard theory is achieved by the use of $\bar{D}$ for all photons represented by internal lines (i.e., for "virtual" photons in the usual usage). One needs to take into account whether energy sources are assumed to be present on either end of the propagation. Thus, within the time-symmetric theory, the use of D_F is really a *practical postulate*, applying to subsets of the universe and/or to postulated boundary conditions consistent with the empirical fact that we observe retarded radiation. It assumes, for example, that the energy source at the earth consists of "free photons" rather than applying a direct-interaction picture in which the energy source photons arise from another current–current interaction and are therefore truly virtual.

The ambiguity surrounding this real vs. virtual distinction arises from the fact that a genuinely "real" photon must have an infinite lifetime according to the uncertainty principle, since its energy is precisely determined at $k^2=0$.[10] But nobody will ever detect such a photon, since any photon's lifetime ends when it is detected, and the detected photon therefore has to be considered a "virtual" photon in that sense. The only way it could truly be "real" would be if it had existed since $t=-\infty$.[11] On the other hand, it is only detected photons that transfer energy; so, as Davies points out, photons that are technically "virtual" can still have physical effects. It is for this reason that PTI eschews this rather misleading "real" vs. "virtual" terminology and speaks instead of offer waves, confirmation waves, and transactions – the latter corresponding to actualized (detected) photons. The latter, which by the "real/virtual" terminology

[10] "Off-shell" behavior applies in principal for any photon that lacks an infinite lifetime; this is expanded in Section 6.3.5.

[11] Of course, this is theoretically possible (even if not consistent with current "Big Bang" cosmology), and could be regarded as the initial condition that provides the thermodynamic arrow, as well as an interesting agreement with the first chapter of Genesis. But the existence of such "primal photons" would not rule out the direct emitter–absorber interaction model upon which TI is based. It would just provide an unambiguous direction for the propagation of positive energy.

would technically have to be called "virtual" since they have finite lifetimes, nevertheless give rise to observable phenomena (e.g., energy transfer). They are contingent on the existence of the offer and confirmation waves that also must be taken into account to obtain accurate predictions (e.g., for scattering cross-sections, decay probabilities, etc.). So in the PTI picture, all these types of photons are real; some are *actualized* – a stronger concept than real – and some are just offers or confirmations. But since they all lead to physical consequences, they are all physically real, even if the offers and confirmations are sub-empirical (recall the discussion at the end of Section 6.2.1; the reality issue is explored further in the next chapter as well).

There is another distinct, but related, issue arising in the time-symmetric approach that should be mentioned. Recall (as noted in Cramer, 1986) that a fully time-symmetric approach leads to two possible physical cases: (i) positive energy propagates forward in time/negative energy propagates backward in time or (ii) positive energy propagates backward in time/negative energy propagates forward in time. Thus the theory underdetermines specific physical reality.[12] We are presented with a kind of "symmetry breaking": we have to choose which theoretical solution applies to our physical reality. In cases discussed above, in which fictitious "free photons" are assumed for convenience, the use of D_F rather than its inverse D_F^* constitutes the choice (i). While this might be seen as grounds to claim that PTI is not "really" time-symmetric, that judgment would not be valid, because it could be argued that what is considered "positive" energy is merely conventional. Either choice would lead to the same empirical phenomena; we would merely have to change the theoretical sign of our energy units.

6.3 PTI applied to QED calculations

6.3.1 Scattering: a standard example

In non-relativistic quantum mechanics, one is dealing with a constant number of particles emitted at some locus and absorbed at another; there are no interactions in which particle type or number can change. However, in the relativistic case, with interactions among various coupling fields, the number and type of quanta are generally in great flux. A typical relativistic process is scattering, in which (in lowest order) two "free" quanta interact through the exchange of another quantum, thereby undergoing changes in their respective energy-momenta p. A specific example is Bhabha (electron–positron) scattering, in which two basic lowest-order

[12] While this might seem a drawback at first glance, the standard theory simply disregards the advanced solutions in an ad hoc manner (which, as noted previously, is inconsistent with the unification of space and time required by relativity). In the time-symmetric theory, the appearance of a fully retarded field can be explained by physical boundary conditions.

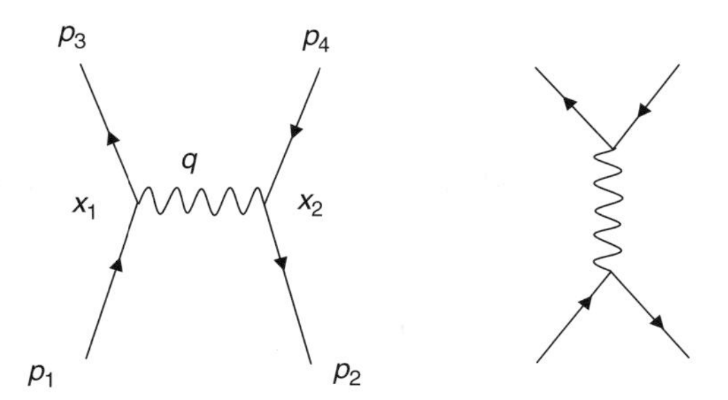

Figure 6.2 Bhabha scattering: the two lowest-order graphs.

processes contribute as effective "offer waves" in that they must be added to obtain a final amplitude for the overall process. The two Feynman diagrams in Figure 6.2 apply in this case.

For conceptual purposes I will discuss a simplified version of this process in which I ignore the spin of the fermions and treat the coupling strength (strength of the field interaction) as a generic quantity g. (The basic points carry over to the detailed treatment with spinors.) In accordance with a common convention, time advances from bottom to top in the diagrams; electron lines are denoted with arrows in the advancing time direction and positron lines with reversed arrows; photon lines are wavy. Key components of the Feynman amplitudes for each process are:

(i) incoming, external, "free" particle lines of momentum p_j, labeled by $\exp\left[-ip_jx_i\right], i,j = 1,2;$[13]
(ii) outgoing, external, "free" particle lines of momentum p_k, labeled by $\exp[ip_kx_i], k = 3,4;$
(iii) coupling amplitudes ig at each vertex;
(iv) an internal "virtual" photon line of (variable) momentum q, labeled by the generic propagator[14]

$$D(x-y) = \int \frac{d^4q}{(2\pi)^4} \frac{ie^{iq(x_1-x_2)}}{q^2} \text{(for } m = 0 \text{ in the photon case)} \tag{6.3}$$

[13] These plane waves are simplified components of the currents appearing in (6.1) and (6.2).

[14] The term "generic" reflects the fact that the denominator here is simply q^2. The different types of propagators involve different prescriptions for the addition of an infinitesimal imaginary quantity, for dealing with the poles corresponding to "real" photons with $q^2 = 0$. However, in actual calculations, one often simply uses this expression. The fact that the generic expression yields accurate predictions can be taken as an indication that the theoretical considerations surrounding the choice of propagator do not have empirical content in the context of micro-processes such as scattering.

To calculate the amplitude applying to the first diagram, these factors are multiplied together and integrated over all spacetime coordinates x_1 and x_2 to give an amplitude M_1 for the first diagram. Specifically:

$$M_1 \propto \int d^4x_1 d^4x_2 e^{-ip_1x_1} e^{-ip_2x_2} (ig) \frac{d^4q}{(2\pi)^4} \frac{ie^{iq(x_1-x_2)}}{q^2} (ig) e^{ip_3x_1} e^{ip_4x_2} \tag{6.4}$$

The integrations over the spacetime coordinates x_i yield delta functions imposing conservation of energy at each vertex (which are conventionally disregarded in subsequent calculations). A similar amplitude analysis applies to the second diagram, giving M_2. Then the two amplitudes for the two diagrams are summed, giving the total amplitude M for this scattering process. M (a complex quantity) is squared to give the probability of this particular scattering process: $\mathrm{P}(p_1, p_2 \rightarrow p_3, p_4) = M^*M$. This is the probability of observing outgoing electron momentum p_3 and positron momentum p_4 given incoming electron and positron momenta p_1 and p_2 respectively. It is interesting to note that the amplitude for the (lowest-order) scattering process is the *sum* of the two diagrams in Figure 6.2, meaning that each is just an offer wave and that the two mutually interfere (see also Figure 6.3).

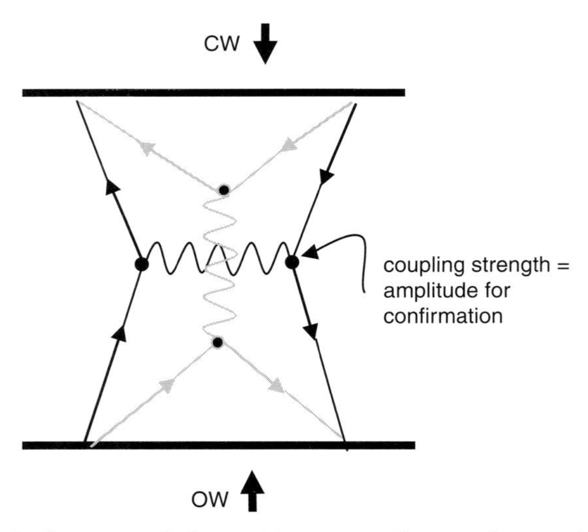

Figure 6.3 Both diagrams of Figure 6.2 are actually superimposed in calculating the lowest-order amplitude for the offer wave corresponding to Bhabha scattering (M_1 shown in black and M_2 shown in gray). Confirmations occur only at the external, outgoing ends. (The CW for this case, a superposition of both scattering OWs, is not shown explicitly in the figure.) Coupling amplitudes at vertices are amplitudes for confirmations that did not, in fact, occur in this process but must still be taken into account in determining the probability for the event.

6.3.2 "Free" particles vs. "virtual" particles

Now, for our purposes, the thing to notice is that, in this very typical analysis, we disregard the history of the incoming particles and the fate of the outgoing particles. They are treated in the computation as "free" particles – particles with infinite lifetimes – whether this is the case or not. And it actually can't be, since we have prepared the incoming particles to have a certain known energy and we detect the outgoing particles to see whether our predictions are accurate. We simply exclude those emission and detection processes from the computation because it's not what we are interested in. We are interested in a prediction *conditioned* on a certain initial state and a certain final state. This illustrates how the process of describing and predicting an isolated aspect of physical reality necessarily introduces an element of distortion in that it misrepresents those aspects not included in the analysis (i.e., misrepresents "virtual" photons – i.e., photons with finite lifetimes – as "real" photons). This is perhaps yet another aspect of the riddle of quantum reality in which one cannot accurately separate what is being observed from the act of observation: the act of observation necessarily distorts, either physically or epistemologically (or both), what is being observed.

6.3.3 The PTI account of scattering

Now, let us see how PTI describes the scattering process described above. There is a two-particle offer wave, an interaction, and a detection/absorption. The actual interaction encompasses all orders[15] – not just the lowest-order interactions depicted here – so the initial offer wave becomes fractally articulated in a way not present in the non-relativistic case. The fractal nature of this process is reflected in the perturbative origin of the S-matrix, which allows for a theoretically unlimited number of finer and more numerous interactions. All possible interactions of a given order, over all possible orders, are superimposed in the relativistic offer wave corresponding to the actual amplitude of the process. (Herein we gain a glimpse of the astounding creative complexity of nature. In practice, only the lowest orders are actually calculated; higher-order calculations are simply too unwieldy, but excellent accuracy is obtained even restricted to these low orders.)[16]

[15] To be precise, all orders up to a natural limit short of the continuum; see note 22.

[16] Adopting a realist view of the perturbative process might be seen as subject to criticism based on theoretical divergences of QFT; i.e., it is often claimed that the virtual particle processes corresponding to terms in the perturbative expansion are "fictitious." But such divergences arise from taking the mathematical limit of zero distances for virtual particle propagation. This limit, which surpasses the Planck length, is likely an unwarranted mathematical idealization. In any case, it should be recalled that spacetime indices really characterize points on the quantum field rather than points in spacetime (Auyang, 1995, p. 48); according to PTI, spacetime emerges only at the level of actualized transactions. Apart from these ontological considerations, progress has been made in discretized field approaches to renormalization such as that pioneered by Kenneth Wilson (lattice gauge theory; cf. Wilson, 1971, 1974, 1975). Another argument against the above criticism of a realist view of QFT's

In the standard approach, this final amplitude is squared to obtain the probability of the corresponding event, but the squaring process has no physical basis – it is simply a mathematical device (the Born Rule). In contrast, according to PTI, the absorption of the offer wave generates a confirmation (the "response of the absorber"), an advanced field. This field can be consistently reinterpreted as a retarded field from the vantage point of an "observer" composed of positive energy and experiencing events in a forward temporal direction. The product of the offer (represented by the amplitude) and the confirmation (represented by the amplitude's complex conjugate) corresponds to the Born Rule.[17] This quantity describes, as in the non-relativistic case, an incipient transaction reflecting the physical weight of the process. In general, other, "rival" processes will generate rival confirmations (for example, the detection of outgoing particles of differing momentum) from different detectors and will have their own incipient transactions. As described in Chapter 4, a symmetry breaking occurs, in which the physical weight functions as the probability of that particular process as it "competes" with other possible processes. The final result of this process is the actualization of a particular scattering event (i.e., a particular set of outgoing momenta) in spacetime.

Thus, upon actualization of a particular incipient transaction, this confirmation *adds* to the offer and provides for the unambiguous propagation of a full-strength, positive-energy field in the $t > 0$ direction and cancellation of advanced components; this is essentially the process discussed by Davies (above), in which the earth–moon energy propagation must be described by D_F rather than by $\bar{D}$.

6.3.4 Internal couplings and confirmation in relativistic PTI

Now we come to an important point. Notice that the internal, unobserved processes involving the creation and absorption of virtual particles are not considered as generating confirmations in relativistic PTI (see Figure 6.3). These are true "internal lines" in which the direction of propagation is undefined; therefore, D_F can be replaced by $\bar{D}$. These must not be confirmed, because if they were, each such confirmation would set up an incipient transaction and the calculation would be a different one (i.e., one would not have a sum of partial amplitudes M_1 and M_2 *before* squaring; squaring corresponds to the confirmation). This situation, involving

perturbative expansion is that formally similar divergences appear in solid state theory, for example in the Kondo effect (Kondo, 1964), but these are not taken as evidence that the underlying physical model should be considered "fictitious."

[17] Technically, by comparison with the standard time-asymmetric theory, the product of the original offer wave component amplitude, $\frac{1}{2}a$, and its complex conjugate, $\frac{1}{2}a^*$, yields an overall factor of $\frac{1}{4}$, but this amounts to a universal factor which has no empirical content since it would apply to all processes and therefore would be unobservable.

summing of several (in principle, an infinite number of)[18] offer wave components to obtain the total offer wave that generates the confirmation), indicates that the field coupling amplitudes, which are not present in the non-relativistic case, represent *the amplitude for a confirmation to be generated.* This is a novel feature of the interpretation appearing only at the relativistic level, in which the number and type of particles can change.

At first glance, this situation may seem odd. If the confirmations at the vertices don't happen, why is there a non-zero amplitude for them to occur? The answer is essentially the same as in the partial amplitudes corresponding to a particle going through either slit in the two-slit experiment. For a particle created at source S, passing a screen with two slits A and B, and being detected at position X on a final screen, the partial amplitudes are

$$\langle \mathrm{X}|\mathrm{A}\rangle\langle \mathrm{A}|\mathrm{S}\rangle \tag{6.5a}$$

$$\langle \mathrm{X}|\mathrm{B}\rangle\langle \mathrm{B}|\mathrm{S}\rangle \tag{6.5b}$$

These must be added together to obtain the correct probability for detection at point X, yet neither generates a confirmation (if both slits are open and there are no detectors at the slits). In each case, no particle was detected at the slit, but the existence of the slit[19] requires that we take it into account. In the same way, the existence of the virtual, intermediate quanta represented in the Feynman diagrams must be taken into account. In quantum mechanics, the unobservable must be accounted for, and it is accounted for in terms of amplitudes (partial offers and partial confirmations), not in terms of probabilities.[20] (The partial confirmations are the advanced wave components from point X on the final screen, through the slits, to the source: $\langle \mathrm{S}|\mathrm{A}\rangle\langle \mathrm{A}|\mathrm{X}\rangle$ and $\langle \mathrm{S}|\mathrm{B}\rangle\langle \mathrm{B}|\mathrm{X}\rangle$.)

6.3.5 Dual role of "current"

I return here to an issue raised obliquely in Chapter 4: the difference between a non-relativistic confirmed current $\varphi^*\varphi$ and the source currents, such as the fermionic current $j^\mu = \bar{\psi}\gamma^\mu\psi$, of quantum field theory. In the Feynman diagram of a scattering amplitude such as that depicted in Figure 6.3, any vertex in which a field

[18] For the present argument, I disregard the issue of renormalization, in which an arbitrary cutoff is implemented in order to avoid self-energy divergences resulting from this apparently infinite regression.

[19] To be more precise in terms of TI, the existence of a large number of absorbers (the slitted screen) which allow only specific OW components to proceed through the experiment.

[20] To be precise, the squaring of the coupling amplitude to obtain the probability of confirmation wave generation is actually the product of the emission coupling amplitude with the annihilation coupling amplitude. That is, a necessary condition for the establishment of an incipient transaction is the generation of *both* an OW and a CW.

quantum is created (such as a photon, the quantum of the electromagnetic field A^{μ}) is always accompanied by both the fermion field operator and its adjoint. This process is what corresponds to the $j^{\mu}A^{\mu}$ in (6.1). But such operator products at vertices are always multiplied by the coupling amplitude (e.g., the "fine structure constant" for QED) and represent possible processes, not actualized events. The possible scattering processes are truly field offer waves; i.e., they are field states characterized by amplitudes, not probabilities. This can be seen from the fact that they arise by the action of field creation operators (multiplied by propagator or "contracted" factors for virtual quanta that are not "free" incoming or outgoing states) on the vacuum state $|0\rangle$. The preceding yields some state of the field, $|\psi\rangle$; in TI terms, an offer wave. The amplitude of this scattering OW is then given by the inner product of the outgoing "free" state $\langle f|$ with the resulting field state (the latter being equivalent to the relevant annihilation field operators acting, from the right, on the adjoint vacuum, $\langle 0|$). This "scattering amplitude" is a complex number, and (after being summed with other applicable processes[21]) it must be squared to obtain the probability of the scattering process characterized by the given incoming and outgoing states. The squaring procedure, $\psi^{\dagger}\psi$, just as in non-relativistic quantum mechanics, corresponds physically to the generation of a confirmation wave at the external (outgoing) end. $\psi^{\dagger}\psi$ is the zeroth component (time component) of the conserved four-current of the associated fields; i.e., $\psi^{\dagger}\psi = \bar{\psi}\gamma^{0}\psi$. The conserved four-current also has the form $j^{\mu} = \bar{\psi}\gamma^{\mu}\psi$, but in this context the fields $\bar{\psi}$ and ψ are not operators, they are states of the incoming and outgoing field, in this case the Dirac field.[22]

6.4 Implications of offer waves as unconfirmed possibilities

As noted in Cramer (1986), this general procedure is not limited to photons. The same principles apply to other types of fields: scalar (Klein–Gordon) particles, Dirac particles (fermions), etc. Note that the need to take confirmations into account for a "real" particle[23] provides a new way to understand the relationship between energy and mass for massive particles, which we explore in this section.[24]

[21] That is, processes of the same level with the same incoming and outgoing states.

[22] We might return here to consider again what Feynman (1985) called the "dippy" process of renormalization. This technique consists in imposing a limit to the fractal infinities of internal processes (specifically, self-interaction loops) in calculating scattering amplitudes. Part of the mystery of renormalization is that different cutoff levels can be chosen, but when computations are appropriately tailored to the chosen cutoff, the same (quite accurate) empirical results are obtained. Fractal processes are known to have a self-similarity in which the different levels of complexity have the same structure, so this picture can account for that aspect of the mystery of renormalization. Under PTI, the fact that these internal processes represent only the *possible* transfer of energy can account for why it is physically valid to "cut" them off; i.e., why the energy associated with them does not have empirical content.

[23] That is, it takes a transacted, confirmed offer wave to result in a detectable transfer of energy from point A to point B.

[24] This section and the following are based on material in Kastner (2011a).

6.4.1 "Real" vs. "virtual" quanta

First, recall the constraint relating rest mass to energy in the usual relativistic expression[25]

$$\omega^2 = k^2 + m_0^2 \tag{6.6}$$

which is the quantum-mechanical version of the usual expression for the relationship between mass, energy, and spatial momentum:

$$E^2 = p^2 + m_0^2 \tag{6.7}$$

Equation (6.6) provides what is termed a *dispersion relation* (a functional relationship between frequency ω and wave number k) for the propagating real wave; this fact will be useful later on. This relation means, in physical terms, that the phase velocity $u = \frac{\omega}{k}$ is not the only velocity associated with the wave; there is also a *group velocity* v_g, given by (refer also to equation (6.6))

$$v_g = \frac{d\omega}{dk} = \frac{k}{\omega} \tag{6.8}$$

The group velocity is the usual particle velocity, i.e., that of a particle with momentum $p = \gamma m_0 v_g$. Note also that $uv_g = \frac{\omega}{k} c^2 \frac{k}{\omega} = c^2$.

However, the masses of virtual particles are not constrained by expression (6.7); indeed, the mass of a virtual particle can be considered as undefined. Such particles are referred to as being "off-shell."[26] Now, the Klein–Gordon equation for real (free) spinless particles with finite rest mass embodies the mass-shell condition, as can be seen from its form (∇ and $\frac{\partial}{\partial t}$ correspond to P and E, respectively):

$$\nabla^2 \varphi - \frac{\partial^2 \varphi}{\partial t^2} = m_0^2 \varphi \tag{6.9}$$

However, a *virtual* KG particle is not constrained by the relationship embodied in (6.9). If virtual particles are identified as offer waves, this supports the idea that it is *confirmations, and resulting actualized transactions*, that enforce the relativistic mass–energy relation describing the classical propagation of energy from one

[25] Here, m_0 is the rest mass of the "particle" associated with the wave. Natural units are used ($c = \hbar = 1$).

[26] The term "virtual particle" is a controversial one. In using it, I want to emphasize that the TI ontology is one of non-local field quanta, not classical corpuscles. "Virtual quanta" are non-local, unobservable objects that should not be thought of as existing in spacetime but rather in "pre-spacetime" (recall Chapter 4). In an ontology in which "real" is viewed as equivalent to "existing in spacetime," virtual quanta are considered "fictitious." As has been argued, I dissent from this position.

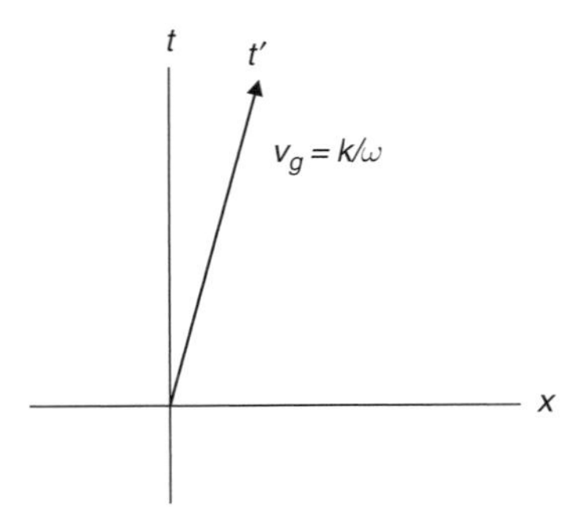

Figure 6.4 The worldline of a real particle.

point to another by way of the group velocity v_g. Returning to the conundrum noted by Davies, discussed above in Section 6.2, a "free" particle with a finite lifetime can readily be identified with an actualized transaction. Unconfirmed offer waves are not constrained by this relation; they do not involve the transfer of empirically measurable energy, and they may therefore be considered to propagate at the phase velocity $u = \frac{\omega}{k}$.

6.4.2 *Offers, transactions, and Minkowski space*

The foregoing sheds an intriguing light on the structure of relativistic spacetime (Minkowski space in the case of special relativity). Consider a real particle, e.g., a KG particle propagating at its group velocity k/ω as in Figure 6.4. This establishes a temporal axis t' for the particle as seen by a stationary observer.

In contrast, a virtual particle, unconstrained by the mass–energy dispersion relation, can be considered to propagate at the phase velocity $u = \frac{\omega}{k}$ of its associated de Broglie offer wave, of the form $\varphi \sim \exp[i(kx - \omega t)]$. If we take into account that, in the particle's rest frame, its de Broglie wave is simply a periodic oscillation of infinite spatial extent (i.e., infinite wavelength),[27] then this oscillation defines an axis of simultaneity for the particle – in other words, a spatial axis. We thus obtain the structure of Minkowski spacetime simply by considering unconfirmed offer waves as specifying a spatial axis, and "real" particles as confirmed and transacted offer waves constrained by the mass–energy dispersion relation (see Figure 6.5).

In this formulation, no energy or mass actually travels at the phase velocity. However, the phase component, which characterizes the offers and confirmations, plays a role in establishing non-local correlations.

[27] This aspect of the de Broglie wave is discussed in de Broglie's dissertation (1924), along with the observation that the group and phase velocities coincide with temporal and spatial axes, respectively (this point is made on p. 12 of the English translation).

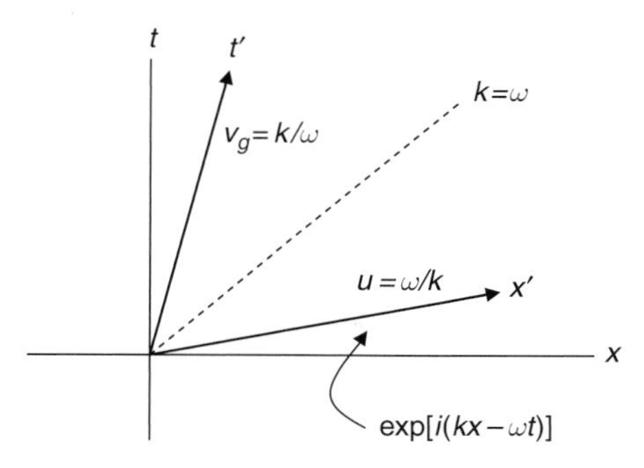

Figure 6.5 The structure of Minkowski spacetime.

de Broglie observed (on p. 12 of the English translation of his dissertation) that the group and phase velocities of his proposed "matter wave" coincided with the temporal and spatial axes of Minkowski spacetime. In Chapter 8, I will explore the idea that spacetime is not just "sitting there" as a substantive container for events, but that it is quanta and their associated de Broglie waves that create spacetime structure through offers and transacted offers.[28]

Shimony (2009) has similarly suggested that relativistic spacetime can be considered as a domain of actuality emergent from a quantum level of possibilities:

> There may indeed be "peaceful coexistence" between Quantum nonlocality and Relativistic locality, but it may have less to do with signaling than with the ontology of the quantum state. Heisenberg's view of the mode of reality of the quantum state was... that it is *potentiality* as contrasted with *actuality*. This distinction is successful in making a number of features of quantum mechanics intuitively plausible — indefiniteness of properties, complementarity, indeterminacy of measurement outcomes, and objective probability. But now something can be added, at least as a conjecture: that the domain governed by relativistic locality is the domain of actuality, while potentialities have *careers* in spacetime (if that word is appropriate) which modify and even violate the restrictions that spacetime structure imposes upon actual events... (2009, section 7, item 2)

Shimony goes on to note the challenges in providing an account of the emergence of actuality from potentiality, which amounts to "collapse." PTI suggests that transactions are the vehicle for this process;[29] and therefore at least part of it must involve processes and entities transcending the spacetime construct.

[28] It is important to keep in mind that an offer wave can receive a confirmation but not result in an actualized transaction. We can refer to this as a "confirmed offer." This corresponds to a null event, i.e., an event that definitely did not occur.

[29] Recall that even if no specific "mechanism" is provided for the actualization of a transaction, TI provides at least a partial solution to the measurement problem since it ends the usual infinite regress by taking into account absorption. A measurement is completed when absorption occurs. Moreover, as suggested above, it is likely misguided to demand a causal, mechanistic account of collapse, since as Shimony suggests, one is dealing with a domain that transcends the causal spacetime realm.

A further comment is in order concerning the puzzle that originally captured de Broglie's interest: the apparent discrepancy between (1) the frequency of oscillation of the phase wave, $f = \gamma m_0 c^2/h$, which is proportional to γ, and (2) the predicted frequency of an oscillation based on a mass m_0 moving at speed v with respect to a stationary observer who is undergoing time dilation and whose frequency, as measured by that observer, is therefore *inversely* proportional to γ (recall the standard moving clock scenario). This discrepancy is resolved by noting that the former, higher frequency corresponds to the superluminal phase wave, and the latter, lower frequency corresponds to an oscillation associated with an actualized system transporting energy. In PTI, the latter would be a phenomenon resulting from actualized transactions. A moving mass that is trackable, i.e., capable of being assigned a determinate trajectory (as in a bubble chamber), will be identified with one or more completed transactions and will therefore be associated with the group velocity and the lower-frequency oscillation.

6.5 Classical limit of the quantum electromagnetic field

It is interesting and instructive to consider how the classical Wheeler–Feynman theory can be seen as a limit of the quantized version. In this section I show how the classical, real electromagnetic field emerges from the domain of complex, pre-empirical offer and confirmation waves that are ontologically distinct from classical fields.

It first needs to be kept in mind that a classical field $E(x,t)$ assumed to propagate in spacetime is replaced by an operator $\hat{E}(x,t)$ in the context of relativistic quantum theory; the latter is a very different entity. It is the transition amplitude of a product of such field operators (actually the vector potential, $\hat{A}(x,t)$[30]) corresponding to two different states of the field (or spacetime points)[31] which then replaces the classical propagating field. That quantity (also known as the Feynman propagator D_F, discussed above, when constructed to ensure that only positive energies are directed toward the future) is now a probability amplitude only, and thus corresponds to the offer wave component of non-relativistic PTI.

Let us now consider how the classical electromagnetic field emerges from the quantum-theoretic electromagnetic field by way of the transactional process. In order to do this, it must first be noted that so-called "coherent states" $|\alpha\rangle$ of quantum fields provide the closest correspondence between these and their classical counterparts. Such states have an indeterminate number of quanta such that annihilation

[30] The electromagnetic field and the electromagnetic vector potential are related by $\vec{E}(x,t) = -\frac{1}{c}\frac{\partial \vec{A}}{\partial t} - \nabla A_t$.

[31] In practice, when the initial and final states are spacetime points, they are just variables of integration. In quantum field theory it is not meaningful to talk about a quantum being created and destroyed at specific spacetime points.

(detection/absorption) of any finite number of quanta does not change the state of the field:

$$|\alpha\rangle = e^{\frac{-|\alpha|^2}{2}} \sum_{n=0,\infty} \frac{\alpha^n}{\sqrt{n!}} |n\rangle \tag{6.10}$$

These states are eigenstates of the field annihilation operator $\hat{a}$; the field in that state does not "know" that it has lost a photon. That is,

$$\hat{a}|\alpha\rangle = \alpha|\alpha\rangle \tag{6.11}$$

so that it has an effectively infinite and constantly replenished supply of photons. The coherent state can be thought of as a "transaction reservoir" analogous to the temperature reservoirs of macroscopic thermodynamics. In the latter theory, the interaction of a system of interest with its environment is modeled as the coupling of the system to a "heat reservoir" of temperature *T*. In this model, exchanges of heat between the reservoir and the system affect the system but have no measurable effect on the reservoir. In the same way, a coherent state is not affected by the detection of finite numbers of photons.

Experiments have been conducted in which a generalized electromagnetic field operator is measured for such a state.[32] Detections of photons in the coherent field state generate a current, and that current is plotted as a function of the phase of the monochromatic source (i.e., a source oscillating at a particular frequency – for example, a laser) (see Figure 6.6). Such a plot reflects the oscillation of the source in that the photons are detected in states of the measured observable (essentially the

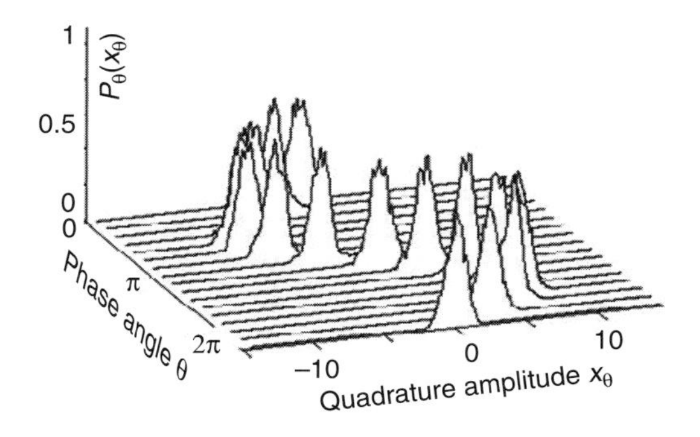

Figure 6.6 Data from photon detections reflecting oscillation of the field source. *Source*: http://en.wikipedia.org/wiki/File:Coherent_state_wavepachert.jpg

[32] See, for example, Breitenbach *et al.* (1997).

electric field amplitude) which oscillate as a function of phase (individual photons do not oscillate, however).

The theoretical difference between the quantum versions of fields (such as the coherent state) and their classical counterparts can be understood in terms of the ontological difference between quantum possibilities (offer and confirmation waves and incipient transactions) and structured sets of actualized transactions. The quantized fields represent the creation or destruction of possibilities, and the classical fields arise from states of the field that sustain very frequent actualized transactions, in which energy is transferred essentially continuously from one object to another. Again, this can be illustrated by the results of experiments with coherent states that "map" the changing electric field in terms of photon detections, each of which is a transaction. For states with small average photon numbers, the field amplitude is small and quantum "noise" is evident (for the coherent state, these are the same random fluctuations found in the vacuum state). As the coherent state comprises larger and larger numbers of photons, the "signal-to-noise ratio" is enhanced and approaches a classical field (see Figure 6.7). Thus, the classical field is the quantum coherent state in the limit of very frequent detections/transactions.

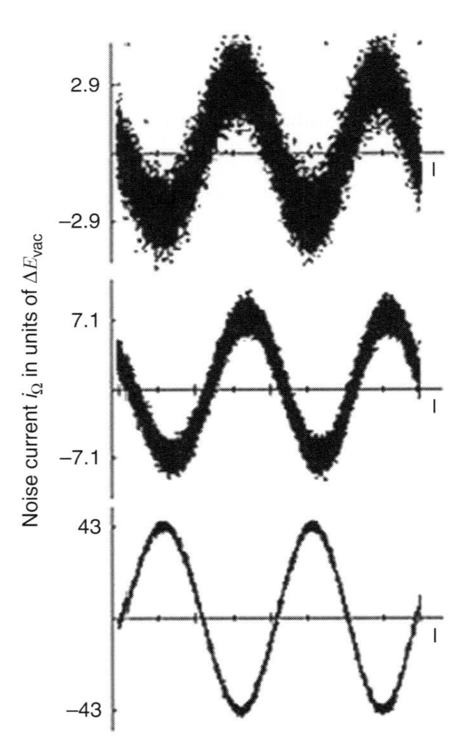

Figure 6.7 Coherent states with increasing average photon number (top to bottom). *Source*: http://en.wikipedia.org/wiki/File:Coherent_noise_compare3.png

It is the classical, continuous detection/transaction limit, in which the field can be thought of as a classical propagating wave, to which the original Wheeler–Feynman theory applies. But it is important to keep in mind the fundamental distinction between a classical field and its quantum counterpart. In this regard, Paul Dirac has observed that:

> Firstly, the light wave is always real, whereas the de Broglie wave associated with a light quantum moving in a definite direction must be taken to involve an imaginary exponential. A more important difference is that their intensities are to be interpreted in different ways. The number of light quanta per unit volume associated with a monochromatic light wave equals the energy per unit volume of the wave divided by the energy $h\nu$ of a single light quantum. On the other hand, a monochromatic de Broglie wave of amplitude a (multiplied into the imaginary exponential factor) must be interpreted as representing a^2 light-quanta per unit volume for all frequencies. (Dirac, 1927, p. 247)

Dirac's comments highlight the ontological distinction between the classical electromagnetic wave and the quantum state (de Broglie wave) for the electromagnetic field. Whereas the classical wave conveys energy through its intensity (the square of its electric field strength), the quantum wave conveys possibility – that is, its square conveys probability in that it represents an incipient transaction whose weight corresponds (in non-relativistic quantum mechanics) to the probability of the corresponding event; or, in the relativistic case of a coherent field state, the *number of quanta* most likely to be actualized. The amplitude of a de Broglie wave for a coherent state with average photon number N is equal to $\sqrt{N}$ (which is proportional to the electric field amplitude for the state); it is a multi-quantum probability amplitude that, when squared, predicts that the most probable number of photons to be detected will be N. Thus, if a coherent state with average photon number 3 were enclosed in a perfectly absorbing box, on examining the box after a time period significantly greater than the inverse frequency of the field (i.e., the period of the oscillation), it would ideally be found to have detected 3 photons.

One could do this by measuring the energy increase of the box, but that is not required; one could imagine a box constructed out of photographic plates that could provide images (dots) of photon absorption. Such images provide a simple numerical answer to the question: "how many photons were actualized?" – and it is to this question that the squared amplitude $(|\alpha|^2)$ of the coherent state $|\alpha\rangle$ applies. In contrast, the squared amplitude of the classical wave addresses the question, "What is the energy associated with the actualized photons?" The *energy* $E = h\nu$ of a particular actualized (detected) photon is frequency-dependent, but the *probable number* of actualized photons is not.

Yet the unity of the two descriptions is still expressed in the fact that it is not the classical field that really conveys energy: rather, it is the *intensity* (squared amplitude) of the field. This can again be traced to the underlying transactional description. A photon does not exist in spacetime unless there is an actualized transaction involving an offer wave and a confirmation wave, which is what is described by the squaring process (Born Rule).[33] Energy can only be conveyed by a detected photon, not by an amplitude (offer wave) only. This fact appears at the classical level and can be seen as a kind of "correspondence principle" between the two descriptions.

6.6 Non-locality in quantum mechanics: PTI vs. rGRWf

GRW approaches were briefly reviewed in Chapter 1. The most recent version of GRW is a proposal by Tumulka, "relativistic GRW flash" or rGRWf, which attempts to provide a relativistically compliant version of that approach, together with a so-called "flash" ontology that provides for specific measurement results without depending on a problematic compression of the wave function.[34] This section argues that PTI does a better job of accommodating relativity.

6.6.1 Gisin's result

Gisin (2010) has recently argued that under certain conditions, and assuming strong causality (i.e., an event can only influence other events in its future light cone), Bell's Theorem will rule out the ability of all hidden variables (whether local or non-local) describable by a covariant probability distribution to reproduce the non-local correlations between spacelike detectors for EPR-type entangled states. Specifically, Gisin considers the usual "Alice and Bob" EPR situation, and defines Alice's and Bob's results α, β respectively as functions F_{AB} [F_{BA}] of their measurement settings $\vec{a},\ \vec{b}$ and the value of some non-local hidden variable λ. The order of the subscripts on F indicates which measurement is first in the frame considered. Thus if Alice measures first, her outcome $\alpha = F_{\mathrm{AB}}(\vec{a}, \lambda)$; if Bob measures first, his outcome $\beta = F_{\mathrm{BA}}(\vec{b}, \lambda)$. Gisin then constructs the analogous function S for the outcome measured second, and notes (assuming time-asymmetric strong causality) that it must also be a function of the measurement setting for the first measurement: i.e., $\beta = S_{\mathrm{AB}}(\vec{b}, \vec{a}, \lambda)$. Analogous expressions are constructed in the frame in which

[33] For those concerned about whether the universe may not be a "light-tight box" as required by traditional DA theories, thus not providing for full future absorption of the OW, it should be noted that confirmations may also be provided by a perfectly reflecting past boundary condition, as proposed in Cramer (1983). This is a type of "absorberless" confirmation in which the advanced wave from the emitter is reflected at $t = 0$ and thereby cancels the remnant advanced wave from the emitter and builds the emitter's retarded OW up to full strength, resulting in an actualized transfer of energy into the infinite future.

[34] One such problem is that a sudden compression of the wave function in the position basis results in an essentially infinite range of energies for the particle.

Bob measures first. Gisin then notes that, if covariance holds, the same λ should characterize the results irrespective of the frame considered, so that we must have

$$\alpha = F_{\mathrm{AB}}(\vec{a},\lambda) = S_{\mathrm{BA}}(\vec{b},\vec{a},\lambda) \tag{6.12a}$$

and

$$\beta = F_{\mathrm{BA}}(\vec{b},\lambda) = S_{\mathrm{AB}}(\vec{a},\vec{b},\lambda) \tag{6.12b}$$

but there is no λ that can satisfy (6.12a,b), since they actually imply that λ is a local variable and these are already ruled out by Bell's Theorem. Thus, Gisin has ruled out the ability of non-local hidden variables to yield a covariant account of actualized outcomes for quantum-correlated spacelike events. This formalizes observations such as Maudlin's (1995) that Bohmian-type "preferred observable" accounts seem to be at odds with relativity.

However, as noted, Gisin's analysis presupposes "strong causality." That is, it specifies which observer's outcome was prior to the other observer's outcome, with the assumption that the second observer's result depends on the setting and outcome of the first observer. Thus, his result does not rule out the ability of time-symmetric approaches to yield a covariant account. Indeed, we will see that PTI can provide all the benefits of Tumulka's GRW "flash ontology" model, "rGRWf" (2006), without being a modification of quantum theory.

6.6.2 Is there really a GRW advantage?

I should first address the claim sometimes made that GRW has an advantage over TI in that the former spells out a particular measurement result while TI's offer/confirmation wave encounter does not (strictly speaking, the latter determines a *basis* for the determinate outcome while not specifying which one occurs[35]). But arguably, this advantage of GRW is only illusory. The GRW outcome is specified by resorting to an ad hoc and physically undefined (in terms of any existing theory) "flash" process. The worst that one can say at present concerning TI (and PTI) is that there is no concrete, causal physical story behind the realization of a particular transaction (outcome) as opposed to a competing "incipient" one, which makes it at least no worse off than GRW in terms of providing concrete physical reasons for a specific measurement result. Meanwhile, TI does give a clear account of the measurement process in terms of absorption, as discussed above. Measurements

[35] It thus gives a physical explanation for the projection postulate of standard QM, as shown in Chapter 3.

are performed by setting up situations in which the particle's offer wave[36] is absorbed at the set of detectors we are interested in, and by discounting runs in which the particle is absorbed (detected) somewhere else.

A common point of confusion concerning TI is the failure to recognize that confirmation waves are generated for *all* components of the offer wave for which absorbers are present, resulting in a weighted set of incipient transactions corresponding to von Neumann's "Process 1" (or the projection postulate). This set of incipient transactions corresponds to an "ignorance"-type mixture, in that measurement has definitely occurred and the uncertainty concerning outcome is epistemic. The realization of a particular transaction out of a set of incipient ones can be seen as a kind of spontaneous symmetry breaking, as discussed in Chapter 4. So it would not be fair to claim, as some have done, that TI is incapable of providing a clear physical account of measurement.

6.6.3 A dilemma re-examined

Tumulka has argued that, in his words, "Either [1] the conventional understanding of relativity is not right, or [2] quantum mechanics is not exact."[37] But this particular dilemma needs to be examined more closely, as horn [1] has more content than is customarily assumed. By [1], Tumulka has in mind the usual assumption that any exact, realist interpretation of quantum theory must involve a preferred inertial frame or "spacetime foliaton." But as noted above, there is something more to be questioned in the "conventional understanding" of relativity: an inappropriately strong time-asymmetric causality constraint. So horn [1] really has two different options: [1a] "there is a preferred frame" or [1b] "causal influences can be time-symmetric." Thus option [1] can be chosen *without* embracing a preferred frame, in the form of [1b]. That is, one can reject the necessity of a preferred frame and argue that what is "not right" about the conventional understanding of relativity is the notion that it mistakenly rules out time-symmetric influences.

Whereas GRW "spontaneous localization" approaches such as Tumulka's "rGRWf," in an effort to avoid the preferred foliation that is assumed to be the only option contained in [1], choose [2] and modify quantum theory in an explicitly ad hoc manner, PTI chooses [1], but not in the sense of [1a] involving a preferred foliation as is usually assumed. Instead, it is noted that relativistic restrictions should be properly considered to apply only to in-principle observable events, and that

[36] The term "particle" is used here for convenience, but recall that in TI there are no "particles" in the usual sense of localized corpuscles pursuing trajectories. As noted by Falkenburg, "The causal particle concept is not just weakened in the subatomic domain, it simply fails . . . the particles are *effects* and their causes are not particles but quantum waves and fields" (2010, p. 329; emphasis added).

[37] Tumulka (2006, p. 352).

sub-empirical causal time symmetry – in the sense of our not being constrained to a choice of which of two events is the "cause" and which the "effect" – should be accepted via option [1b].

Indeed, a similar relaxation of strong causation is just what Tumulka adopts in order to argue that the non-local correlations arising between spacelike separated flash events in his model do not violate covariance. He remarks: "An interesting feature of this model's way of reconciling nonlocality with relativity is that the superluminal influences do not have a direction; in other words, it is not defined which of two events influenced the other."[38] Note that, since these are spacelike separated events, there is a frame in which one is first and a different frame in which the other is first, so one could argue that there can be time-reversed causal effects in one frame or the other, depending on which event is arbitrarily considered the "cause" and which the "effect." (One might object here that Tumulka addresses this by saying that no such causal order exists, but that is precisely the case in PTI as well.) So we see the relativistic version of GRW already heading in the direction of time symmetry, or at least toward weakening the overly strong "causality" assumption so often presumed in the literature.

Under PTI, sets of possible transactions (whose weights, interpreted as probabilistic propensities, are reflected in the Born Rule) provide a covariant, time-symmetric distribution of possible spacetime events. Moreover, there is nothing about the sets of actualized events in PTI that can be seen as non-covariant, as in the actualized events discussed by Gisin. This is because, under PTI, it is not assumed that the events (Alice and Bob's outcomes) had a strict temporal causal order. Gisin's observation regarding the non-covariance of actualized events does not apply to sets of actualized events in PTI, since all events are dependent on both the emitter's "offer wave" and the absorber(s') "confirmation wave(s)." Just as in Tumulka's account of his non-locally correlated flashes, there is no need (nor would it be appropriate) to define which of a set of spacelike separated events is the "cause" and which is the "effect" of a particular outcome. The emitter and absorber(s) participate equally and symmetrically in the transaction leading to the outcome(s). Thus, actualized transactions play the part of the "flashes" in Tumulka's model, but without the necessity of modifying the dynamics of quantum theory. While Tumulka has opted for a modification of quantum theory in order to avoid a preferred frame – our [1a] above – he has also made use of [1b] which, in view of the time-symmetric alternative of PTI, obviates the need for modifying quantum theory in the first place.

[38] Ibid.

6.7 The apparent conflict between "collapse" and relativity

Finally, I address the conundrum of "collapse" and its apparent clash with relativistic precepts. In a definitive paper, Aharonov and Albert (1981) discuss several conceptual difficulties surrounding "collapse" of quantum states, which I review below.

6.7.1 Instantaneous collapse violates relativity

If a measurement is made at $t=0$ and collapse is instantaneous[39] (see Figure 6.8), this is manifestly non-covariant, since the same collapse will not be instantaneous for a different inertial observer. One can try to address this by considering collapse as occurring along the past light cone (Figure 6.9);[40] but in the usual story where measurement is considered to occur at a single spacetime point, we run into difficulty as follows.

The position measurement is assumed to occur at $(0, 0)$. If, however, we conduct a momentum measurement at $t=0-\varepsilon$ (ε small), this should confirm the prepared momentum eigenstate as shown in Figure 6.8. Yet Figure 6.9, in which "collapse" occurs along the backward light cone, clearly shows that the particle is *not* in a

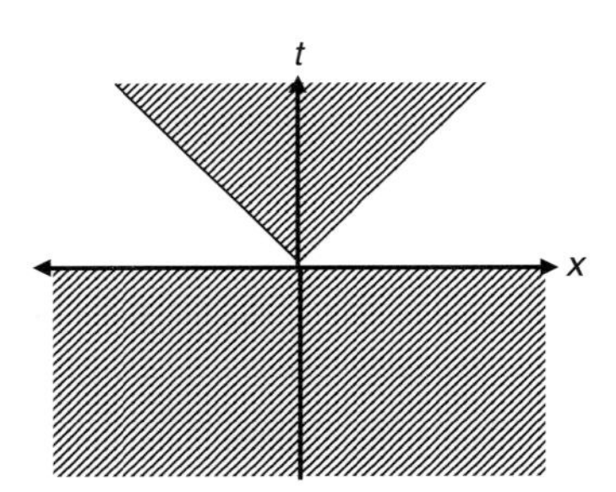

Figure 6.8 Instantaneous collapse.

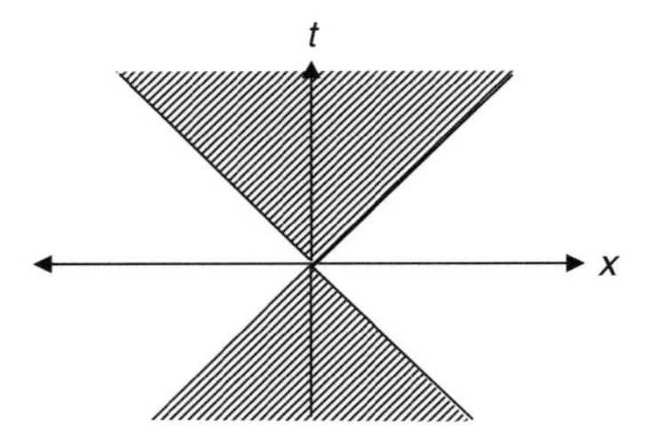

Figure 6.9 Collapse along the past light cone.

[39] By "collapse being instantaneous," I mean that direct causal effects of the collapse "travel" at infinite speed.
[40] This proposal was explored by Hellwig and Kraus (1970).

momentum eigenstate for $t < 0$.[41] AA show, contrary to earlier claims by Landau and Peierls (1931), that such a measurement could indeed be conducted. Therefore, they argue that collapse really does have to be instantaneous. Their overall conclusion is that quantum states are not covariant objects (meaning that they apparently are not consistent with the theory of relativity). This conclusion can be seen as consistent with PTI, insofar as relativity is viewed under this interpretation as applying only to the empirical level of actualized transactions. "Quantum states" in PTI are just offer waves, which in themselves do not result in empirical phenomena associated with specific properties in spacetime, and as such are not subject to relativistic restrictions.

6.7.2 Momentum eigenstates are non-local

Moreover, clearly a momentum eigenstate is a non-local object, which (as the discussion by AA notes) in itself seems to violate relativity. However, if relativity properly applies strictly to the empirical level – an eminently reasonable assumption consistent with all observable phenomena – the non-covariance of offer waves as sub-empirical entities does not have to be seen as a theoretical or methodological problem. Relativity was famously invented to conform to specifically empirical constraints; momentum eigenstates are manifestly sub-empirical. In addition, the ontology of "collapse" is sub-empirical in PTI, as will be shown below.

Under PTI, a particle prepared in a momentum eigenstate is a non-local offer wave that generates confirmation waves from all possible absorbers (e.g., atoms in their ground states). This sets up an enormous number of incipient transactions corresponding to all accessible absorbers. The collapse (realization of a particular transaction) is an interaction between the *non-local* momentum offer wave $|p\rangle$ and a particular confirmation wave $\langle x_k|$. It is a binary interaction between two states. This process is indicated in Figure 6.10. What propagates along the backward light cone is a confirmation wave – not a quantum state, which under PTI is an offer wave that does not correspond to a particular empirical property at a particular time. Put differently, outcomes are not identified with states; they are identified strictly with projection operators $|x\rangle\langle x|$ that, under PTI, refer unambiguously to an encounter

[41] Marchildon (2008) argues that the momentum measurement at $t = 0 - \varepsilon$ could be conducted by several carefully correlated local measurements, and this would collapse the momentum state along backward light cones corresponding to the points of those measurements. He constructs a possible consistent solution in which the position measurement does not contradict the momentum state, but this depends on assuming that the position measurement's influence along the backward light cone stops at the backward light cone of the momentum measurements. It is not clear whether nature would actually allow the influence of that position measurement to stop at this point. If the influence of the position measurement continues past infinity, there is still a contradiction between the two states.

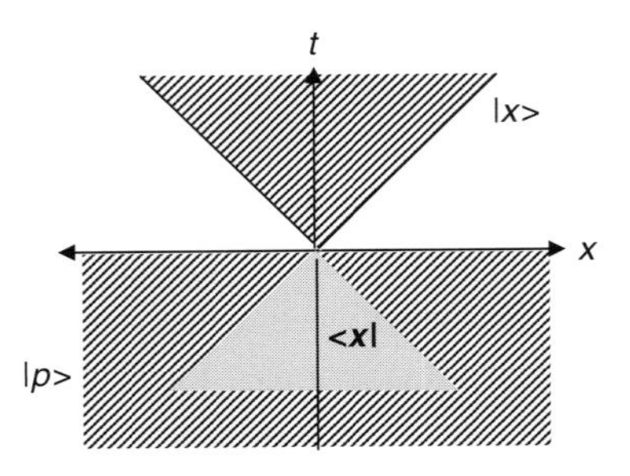

Figure 6.10 The advanced confirmation wave $\langle x|$ corresponds to the new position eigenstate offer wave $|x\rangle$ created at $t = 0$ and does not contradict the preceding momentum eigenstate offer wave $|p\rangle$.

between an OW and a CW. (In standard QM these quantities lack a specific physical reference.)

If the momentum measurement is performed at $t=0 - \varepsilon$, this will generate a confirmation wave $\langle p|$ that, based on the offer wave $|p\rangle$, will with certainty result in the realized transaction corresponding to $|p\rangle\langle p|$. An interval $t=\varepsilon$ later, the position measurement yields a result $|x_k\rangle\langle x_k|$. This *outcome* is not something that travels along the backward light cone; only the confirmation $\langle x_k|$ travels along the backward light cone[42] and it therefore does not contradict the original offer $|p\rangle$, which is a different entity. Moreover, if the universe is a "light-tight box" – meaning that no free fields exist in the infinite past or infinite future – then the backward-propagating $\langle x_k|$ is canceled and there is no residual advanced entity at all. In any case, the crucial point is that, in PTI, an outcome is not identified with a state; rather, it results from an interaction between *two* objects, an offer and a confirmation, and collapse occurs with respect to those two states. Thus measurement in PTI is binary rather than unary.

6.7.3 Collapse is not a spacetime process

It should be noted further that, if we were dealing with an initial superposition of momentum eigenstates, say $|p_1\rangle$ and $|p_2\rangle$, followed by a momentum measurement, then clearly a collapse to either of these states would be non-local in any case, since momentum states are completely non-local objects. This underscores the futility of trying to locate "where in spacetime" collapse occurs. Collapse is not located somewhere in spacetime. If spacetime is the empirical arena, collapse is a completely sub-empirical process.

[42] This is assuming that offers and confirmations travel with the speed of light, which may not be the case for massive particles (OW for the latter may propagate superluminally). This is addressed in Chapter 8.

Under standard approaches to QM, we can usually get away with identifying outcomes with states for the following reason. If we perform a non-demolition measurement that detects a system at location X at time t, but allow it to propagate further, what propagates is an offer wave, which under standard QM is identified with the outcome. That is, the state $|X\rangle$ is typically considered a necessary and sufficient condition for the possession by the system of the property X, but under PTI it is not. Under PTI, an offer wave $|X\rangle$ is a necessary but not sufficient condition for attributing a particular property X to the system.

6.8 Methodological considerations

An interesting aspect of the Davies (and Wheeler–Feynman) time-symmetric theory is that it is theoretically falsifiable. Falsification of the theory would involve being able to show that, given known universal boundary conditions, a particle that should radiate is not, in fact, radiating. While such tests would be extremely difficult (if not practically impossible) to carry out, the fact that the theory is in-principle falsifiable underscores its methodological superiority to the standard ad hoc approach of simply discarding the advanced solutions. Nevertheless, it should also be noted that PTI itself does not stand or fall with the explicit time-symmetric theories of Wheeler and Feynman or Davies. It can be maintained as an interpretational framework even under the standard theory if one interprets the squaring process as representing the existence of advanced confirmation waves (at the quantum, not classical level) which are not usually taken into account. (See, e.g., Chiatti, 1995). PTI is presented here in terms of specific time-symmetric theories because that is the most natural theoretical basis for the interpretation, even if not a crucial one.

7

The metaphysics of possibility in PTI

> All the world's a stage,
> And all the men and women merely players:
> They have their exits and their entrances;
> And one man in his time plays many parts . . .
>
> *Shakespeare,* As You Like It

PTI is a realist interpretation which, in its strong form, takes the physical referent for quantum states[1] to be ontologically real possibilities existing in a pre-spacetime realm, where the latter is described by Hilbert space (or – more accurately – Fock space, accommodating the relativistic domain). These possibilities are taken as real because they are physically efficacious, leading indeterministically to transactions which give rise to the empirical events of the spacetime theater. PTI can also be considered in a weaker, agnostic, "structural realist" version, in which the Hilbert space structure of the theory is taken as referring to some structure in the real world without specifying what that structure is. (I specifically address the structural realism aspect in Section 7.6.) PTI in its strong form is very different from the traditional "possibilist realism" or "modal realism" pioneered by David Lewis. In order to make this distinction clear, I first briefly review the traditional account.

7.1 Traditional formulations of the notion of possibility

As noted in Chapter 1, David Lewis pioneered realism about possibilities in a comprehensive and sustained philosophical examination of entities he termed "possible worlds" (Lewis, 1986). In Lewis' formulation, possible worlds are the

[1] The term "semantic realism" is often used to denote the idea that theoretical terms refer to specific physical entities, the position I advocate herein concerning quantum theory. In contrast, "epistemic realism" denotes the idea that we have good reason to believe a theory's claims. I consider a stance of epistemic realism about quantum theory as relatively uncontroversial, so I do not address it here.

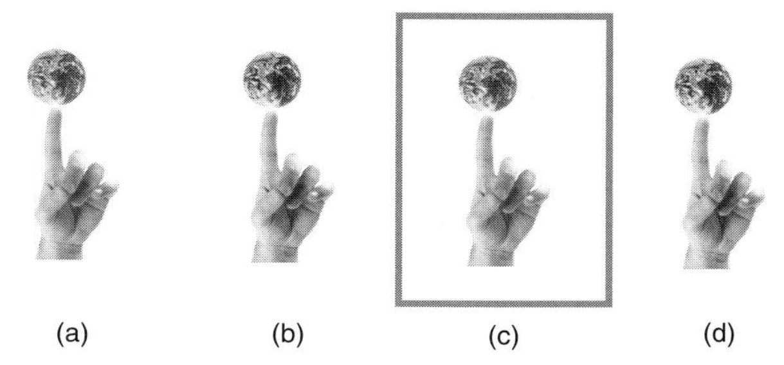

Figure 7.1 A set of "possible worlds" in traditional Lewisian possibilist realism. Worlds (a), (b), and (d) are possible worlds; the "actual world" (c) (in rectangle) is defined only relative to an observer. Each world is considered to be a complete, universal set of events.

same sorts of entities as our own world. They are states of affairs that could conceivably occur, but which differ from the set of events in the actual (experienced) world. According to Lewis, these worlds are every bit as real as the actual world; the only difference is that the actual world is the one we happen to inhabit. Thus, in this theory, "actual" is *indexical*, meaning that it is a matter of perspective, not of kind or nature. Figure 7.1 illustrates this relationship schematically between Lewisian possible worlds and the actual world.

The Lewisian formulation is readily applicable to "many worlds"-type interpretations, in which each measurement event[2] causes a "branching" or copying of a particular world or collection of objects. However, PTI's proposed dynamic possibilities are fundamentally different from those of the Lewisian picture, as will be discussed in the next section.

7.2 The PTI formulation: possibility as physically real potentiality

As noted above, Lewisian possible worlds are just alternative universal states of affairs, and are no different in their basic nature from the actual world. In contrast, the dynamical possibilities referred to by state vectors in PTI are Heisenbergian "potentia," which are less real than events in the actual world, yet more real than mere thoughts or imaginings or conceivable events. This relationship is illustrated in Figure 7.2.[3] In contrast, as noted in Chapter 1, traditional approaches to

[2] Recall that, as discussed in Fields (2011), the notion of a "measurement event" is ill-defined in Everettian interpretations because it requires dividing the physical objects under study into those which constitute the "measured system" and the "measuring apparatus." Such a specification is non-unique and therefore requires reference to an external observer or arbitrary choice.

[3] Actually, mental activity *could* be considered real as well in that it could be based on quantum possibilities; this remains an interesting metaphysical question, but it is not crucial for PTI.

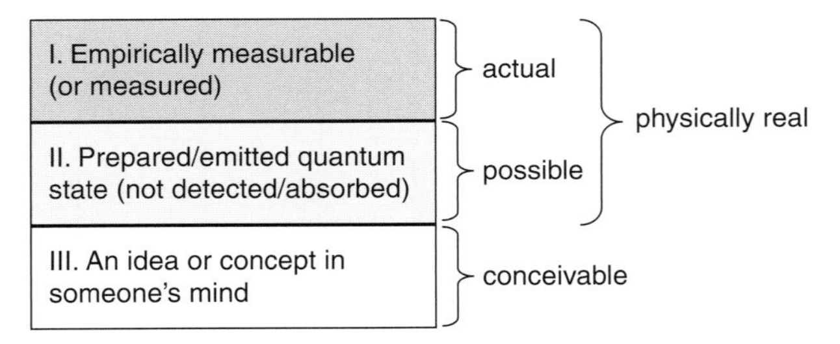

Figure 7.2 Quantum entities are less real than empirically measurable events, but more real than thoughts or merely conceivable situations.

measurement in quantum theory inevitably end up needing to invoke an "observing consciousness" in order to "collapse" the wave function (or state vector) and bring about a determinate outcome, necessitating speculative forays into psycho-physical parallelism. Thus, PTI is actually *less radical* than these much more common approaches because it does not need to invoke mental substance in order to address what certainly started out as a purely physical, scientific question about material objects.

Under PTI, the realist use of the term "possible" or "potential" refers to physical possibilities; that is, entities which can directly give rise to specific observable physical phenomena based on an actualized transaction.[4] This is distinct from the common usage of the term "possible" or "possibility" to denote a situation or state of affairs which is merely conceivable or consistent with physical law. So, in general, "possibilities" in PTI are entities underlying specific *individual* events rather than collective, universal sets of events such as the worlds in Figure 7.1. In more technical terms, the possibilities underlying, for example, the detection of a photon at point X on a photographic plate are the offer wave components constituting the path integral in Feynman's "sum over paths" (recall Chapter 4).

Specific examples of each metaphysical category illustrated in Figure 7.2 are:

I. A detector click.
II. A spin ½ atom prepared in a state of "up along x."
III. "That possible fat man in the doorway."[5]

[4] This is very similar to, indeed perhaps the same as, Teller's proposal (Teller, 1997, 2002) that (however negatively stated in the words of Frigg, 2005, p. 512), the quantum field "has only something like structural efficacy, meaning that it does no more than [specify] the structure of physically possible occurrences."

[5] This is a reference to a famous 1948 paper by Quine, "On what there is," in which he criticizes traditional possibilist realism because of the apparent proliferation of any conceivable entity in a "slum of possibles" that is a "breeding ground for disorderly elements." (Reprinted in Quine, 1953.)

7.3 Offer waves, as potentia, are not individuals

A significant component of the literature in philosophy of quantum theory is addressed to understanding the metaphysical nature of quantum systems such as electrons in the following sense: are they individuals, i.e., do they have some "essence" above and beyond the usual dynamical attributes such as momentum, spin, and (in traditional approaches) spacetime location, etc.? In the PTI picture, the answer to this question is an unequivocal "no."[6] This is because the PTI (as well as original TI) ontology has no "particles" to whom one could even begin to attribute individualized "essences" or identities. In Section 7.3.1 we will see that a direct consequence of the non-existence of particles is that quantum states are restricted in their mathematical form to be either symmetric (meaning unchanged under an exchange of subsystem labels) or antisymmetric (meaning changing only by a sign under an exchange of subsystem labels), and must therefore be either bosons or fermions. (The latter feature of the quantum mechanics of multi-particle systems is sometimes viewed as a curious fact in need of explanation).[7] In Section 7.3.2 I will discuss the apparent dependence of particle number on an observer's state of motion, which also suggests that the notion of particle is not fundamental.

7.3.1 Wave function symmetry related to non-existence of particles

First, recall that standard quantum mechanics assigns to a quantum emitted from a specific location in the laboratory, at some time $t=0$, a Gaussian wave function[8] depending on the amount of time elapsed since its emission. Such a wave function is illustrated schematically in Figure 7.3(a). Now, suppose two quanta of the same type are emitted at $t=0$ (say, both electrons). If sufficient time has elapsed, the wave function for the two quanta looks like Figure 7.3(b): that is, there is significant overlap (cross-hatched region). The usual way of discussing this is to say that there is no way to know which particle is described by which wave function, and therefore one has to assume that the particles are indistinguishable, where their indistinguishability is contingent on the fact that wave functions can overlap. However, in the TI/PTI ontology, there are no "particles" associated with either wave function, independently of whether or not the wave functions overlap. This leads to a different, but arguably stronger, demonstration of the fact that quantum states must be either symmetric or antisymmetric.

[6] Thus I agree with Teller's view (1997) that quanta lack "primitive thisness."

[7] In particular, O. W. Greenberg has explored the idea of "parastatistics" in which the quanta are neither bosons nor fermions.

[8] Again, offer waves are not restricted to being wave functions, which are committed to a particular basis (namely the position basis); but this is probably the most familiar and intuitively easy way to conceptualize the issue under study.

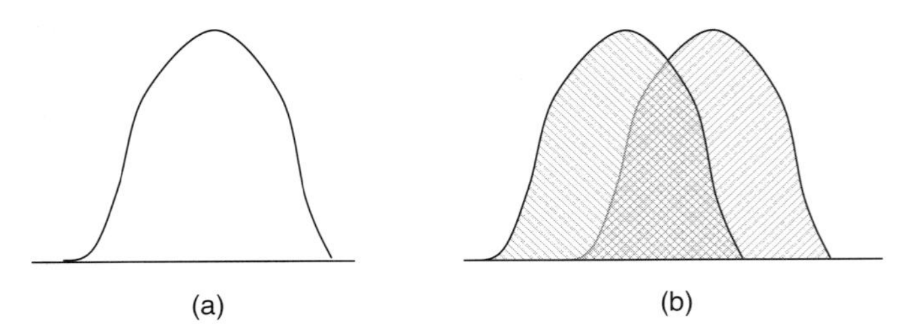

Figure 7.3 (a) The Gaussian wave function of a free quantum. (b) Overlapping wave functions of two free quanta.

The usual way of arguing that quantum states must be either symmetric or antisymmetric is by demanding that observable quantities (such as probabilities of detection) be invariant under a change of particle labels. For example, consider (as in Eisberg and Resnick, 1974) two particles in a 1-dimensional box of side length a, one of them occupying the ground state $G(x_1)$ and the other occupying the first excited state $F(x_2)$, where x_1 and x_2 denote the location of each of the particles. (The two functions G and F have very different dependences on spatial location x.[9]) Now consider a non-symmetrized two-particle wave function such as $\Psi(x_1, x_2) = G(x_1)F(x_2)$. The probability density will be

$$\mathrm{P}(x_1, x_2) = \Psi^*(x_1, x_2)\Psi(x_1, x_2) = G^*(x_1)F^*(x_2)G(x_1)F(x_2) \tag{7.1}$$

But if we transpose the particle labels, then we get

$$\mathrm{P}(x_2, x_1) = \Psi^*(x_2, x_1)\Psi(x_2, x_1) = G^*(x_2)F^*(x_1)G(x_2)F(x_1) \tag{7.2}$$

In equations (7.1) and (7.2) we have the functions G and F and their complex conjugates evaluated at different points x_i, so the probability densities $\mathrm{P}(x_1, x_2)$ and $\mathrm{P}(x_2, x_1)$ are not necessarily equal. In order to make them equal, we have to construct either the symmetric wave function Ψ_s or the antisymmetric wave function Ψ_A,

$$\Psi_S(x_1, x_2) = \frac{1}{\sqrt{2}}\left[G(x_1)F(x_2) + G(x_2)F(x_1)\right]$$

$$\Psi_A(x_1, x_2) = \frac{1}{\sqrt{2}}\left[G(x_1)F(x_2) - G(x_2)F(x_1)\right]$$

[9] For this example, they are $G(x) \sim \cos\frac{\pi x}{a}$ and $F(x) \sim \sin\frac{2\pi x}{a}$.

Thus, to review, the usual argument demands that empirically observable quantities such as the probability density be invariant under a transposing of particle labels based on the premise that quantum objects are "indistinguishable." The latter premise is arrived at because of an argument such as "wave function overlap makes it impossible to tell which particle is associated with which wave function."

Now suppose there are no particles at all. Then there is no auxiliary entity to associate with a wave function which could be "labeled," and which therefore could be addressed by the above sort of argument. But we can arrive at the need for symmetrization more directly as follows. Consider equations (7.1) and (7.2). If there are no particles whose labels could be transposed, the only way to make these two expressions equal is to demand that $x_1 = x_2$. But if we do that, the resulting wave function only refers to *one* quantum. In the absence of auxiliary (labelable) quantum entities, the only way we can enforce the fact that there are two quanta is to provide two distinct arguments x_1 and x_2. Then the arguments *don't label anything*, but they are required in order to distinguish between a wave function for only one quantum and a wave function for two quanta. If they don't label anything, then there can be no physically appropriate meaning in an expression like $G(x_1)F(x_2)$, which implies a difference between the two arguments of the functions G and F. The mathematical expression of the fact that *there is no physical difference between the two arguments* is precisely the set of symmetric and antisymmetric wave functions above. Thus, *the observed fact that nature has only bosons (represented by symmetric states) and fermions (represented by antisymmetric states) can be arrived at simply by assuming that there are actually no "particles" (or individuals) meriting labels of any kind*. Again, we return to the idea that the fundamental ontological reality is that of non-localized fields and their excitations. The new feature proposed in PTI is that these fields represent possibilities for transactions, the latter corresponding to specific observable events.

7.3.2 The puzzle of "Rindler quanta"

An ongoing discussion in the literature concerns so-called "Rindler quanta." These are excited states of the field which are only seen by an observer in a state of constant acceleration with respect to an inertial observer who (in contrast) sees a vacuum (unexcited) state of the field. The field excitations have the form of a "thermal bath," which is similar to the coherent state discussed in Chapter 6 in the sense that it contains an indefinite number of quanta. An accelerating detector registers quanta from this "thermal bath" which does not seem to be present to the inertial observer. The phenomenon of Rindler quanta has serious implications for the question of the "reality" of quanta, since it seems to tell us that not only the properties of quanta but even *whether or not there are any quanta* is a purely "contextual" matter – i.e., dependent on the observer and what types of measurements he/she chooses to make.[10]

From the PTI standpoint, the problem evaporates. There are no independently existing "quanta" in either case; there are simply possible transactions. Rindler phenomena tell us that an accelerating observer (modeled by a simple accelerating detector in the literature on this topic) simply has a different perspective on the relevant transactional opportunities than does an inertial observer. In particular, to an inertial observer, the accelerating detector *emits* quanta, which are simply related to the energy of its acceleration. Thus, a transaction appearing to an inertial observer as a quantum emitted by the accelerating detector and received by an inertial detector is seen by the accelerating observer as a quantum emitted by the field and received by the accelerating detector. In both cases, a transaction occurs; it is simply interpreted differently by the different observers. The two observers define their "field vacuum state" differently;[11] they experience the very same transaction, but seen from different perspectives based on their differing reference vacuum states. Since transactions, and the possibilities leading to them, are the fundamental ontological entities in TI – rather than quanta – TI has no trouble accounting for the phenomenon of Rindler quanta.

7.4 The macroscopic world in PTI

In this section, I consider macroscopic objects and the everyday level of experience in the transactional picture.

7.4.1 Macroscopic objects are based on networks of transactions

I said in the previous section that there are no individual "particles," just field excitations – Heisenbergian "potentia" – that can lead to observable events via actualized transactions. Here I wish to address the question: what is it about transactions that make events "observable"?

First, recall that it is only through actualized transactions that conserved physical quantities (such as energy, momentum, and angular momentum) can be transferred. Such transfers occur between emitters and absorbers which are also field currents (recall Chapter 6). Thus the supporting entities and structures for actualized transactions are generally only potentia themselves. The realizing of phenomena is a kind

[10] Cf. "Are Rindler quanta real?" (Clifton and Halvorsen, 2000), http://philsci-archive.pitt.edu/73/1/rindler.pdf.

[11] The Rindler vacuum state actually has *negative* stress–energy density in an amount exactly balanced by the stress–energy density in the "thermal bath" of Rindler quanta. Cf. DeWitt (2003, pp. 608–18). Thus the Rindler vacuum and the thermal bath together are equivalent to the Minkowski (inertial) vacuum in terms of energy.

 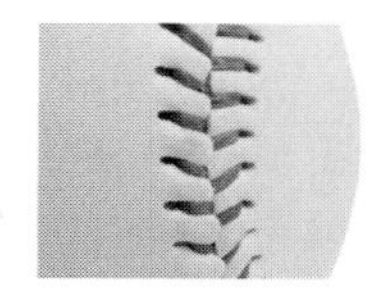  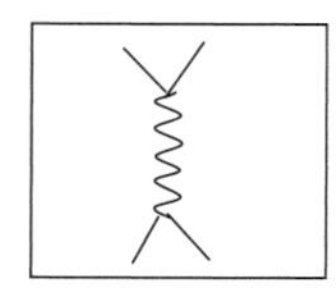

Figure 7.4 Zooming in on a baseball.

of "bootstrapping" process in which actualized events are rooted in unactualized possibilities.

For a specific illustration, consider a baseball, depicted in Figure 7.4 as we "zoom in" to view it on smaller and smaller scales. The third square represents molecular constituents; the fourth square, a Feynman diagram, represents interactions among subatomic constituents both within molecules (intramolecular forces) and between molecules (intermolecular forces). Bound systems such as atoms are only offer waves, but they can (and do) continually emit and absorb photons and other subatomic quanta. Those emitted quanta are absorbed by, for example, our sense organs, setting up enormous numbers of transactions transferring energy between ourselves and the atomic constituents of the baseball. The energy transfers effect changes in our brain, providing for our perception of the baseball.

Thus, in the TI (and PTI) picture, a necessary feature and key component of any observation of a system is absorption of offer waves and corresponding generation of confirmation waves. We can go further and make a general interpretational identification of absorption with observation in a way not available to traditional interpretations of quantum theory: absorption is the way the universe "observes itself" and makes things happen. This identification is possible because under TI, absorption plays an equal role with emission in the dynamics of an event. In contrast, traditional interpretations take emission as the entire dynamical story and then cannot account for why observations seem to have such a special role in the theory. As Feynman tells us, we should sum the amplitudes over "unobserved" intermediate stages of an event to get a total amplitude for a final "observed" event, and then take the square of that. Why should we square that amplitude, and why should nature care whether we "observe" or not in this algorithm? *The only way that nature could know or care would be because something physical really happens in such "observations," and the only possible physical process accompanying an "observation" is absorption.* Under traditional interpretations which neglect absorption, the above apparently inexplicable procedure leads us into an impenetrable thicket of anthropomorphic considerations of the supposed effect of a mental substance – "consciousness" – on a physical substance, namely a quantum system. In Feynman's words: "*Do not keep saying to yourself, if you can possibly avoid it,*

'But how can it be like that?' because you will get 'down the drain', into a blind alley from which nobody has escaped. Nobody knows how it can be like that."[12] I suggest that an escape route from the "blind alley" is available; the price (or dividend, depending on one's point of view) is taking absorption into account as a real dynamical process and embracing the implications for our world view which are explored in this and the next chapter.

7.4.2 Macroscopic observation as primarily intersubjective

Next, let's consider a prototypical observation: once again, the two-slit experiment. Let's assume that the quanta under study are monochromatic (single-frequency) photons originating from a laser. In setting up the laser and the two screens, we handle macroscopic materials such as photographic plates. All of these actions consist of molecular-level transactions between enormous numbers of atoms and between some of the surface atoms and our hands. Energy is transferred via these transactions from those emitters to absorbers on our bodies; that energy serves as input for additional emissions between our sense organs and absorbers in our nerves, and so on, culminating in transfers of energy to our brains.[13] Brain changes make possible our perception that "something happened" (recall, from Chapter 2, Descartes' argument that it is not possible to observe anything that does not produce a perceptible change). But exactly *what happened* can vary considerably, depending on the specific transactions being actualized. A transaction between the photographic plate and my retina will not be the same as the transaction between another part of the plate and someone else's retina, but the laws of physics[14] ensure that all those many transactions are coordinated such that a coherent set of phenomena are created.

The point is that a macroscopic "observed event" is generally the product of an enormous number of transactions, even for only one observer. If one wishes to have one's observation corroborated, more transactions are required as another set of eyes, hands, etc. are introduced. These comprise a different set of absorbers, and the emitters may well be different as well. The transactions occurring for the second observer are not the same as those occurring for the first observer. For there to be corroboration, the two observers have to agree on macroscopic facts such as "There is a dark spot at position $x = 50$," which can be instantiated by a large number of different sets of microscopic transactions. The process of corroboration is thus one

[12] The Messenger Lectures, 1964, MIT.

[13] This description is not meant to be physiologically rigorous; it is merely an indication of how energy transfers via transactions ultimately result in brain changes.

[14] For example, conservation of physical quantities corresponding to the symmetries of the system and compliance with such laws as the principle of least action.

of comparing the transaction-based perceptions of two (or more) different observers and deciding whether they represent the same macroscopic event. *But the event itself can be no more than the sets of transactions taken as constituting it.* It is always definable only in terms of the subjective or intersubjective experiences of an observer or observers.

The above should not be taken as a reversion to mere subjectivism,[15] since for any individual transaction between emitter and absorber, there *is* an objective matter of fact concerning which transaction was actualized. Furthermore, there are certainly experiments in which an individual actualized transaction can be amplified to the macroscopic level, as in detection by a photomultiplier. But even in the case of amplification of a single transaction to the observable level, the type of event observed depends on what absorbers are present for the emitted quantum. In general, ordinary events are collections of enormous numbers of transactions, with different sets of transactions for different observers.

7.4.3 Implications for the realism/antirealism debate

The PTI account of observation provides for a synthesis of the longstanding "realism/antirealism" dichotomy in that both doctrines can be seen as conveying a partial truth. Let us first briefly review these doctrines.

The doctrine of realism spans many forms, from the "naïve realism" most of us grow up believing, to much more sophisticated forms, including "scientific realism," that have evolved in philosophical debate. For our purposes, we can make do with a definition from the *Stanford Encyclopedia of Philosophy*: "Metaphysically, [scientific] realism is committed to the mind-independent existence of the world investigated by the sciences."[16] The world and the entities in it are assumed by the scientific realist to exist independently of our minds, perceptions, and knowledge. The objects in our world are considered as possessing definite properties, which we can come to know without fundamentally disturbing or changing those basic properties.

Antirealism denies this view; it asserts that objects of knowledge are dependent on (or constituted by) some form of subjectivity or mental substance. For example, the philosopher and Irish cleric George Berkeley famously asserted – and ably defended – the doctrine *esse est percipi* (to be is to be perceived), and concluded that all objects are ultimately ideas in the mind of God.[17] The work of Immanuel Kant

[15] Subjectivism is the view that knowledge can only be about experiences of a perceiving subject and not about any genuine object external to the subject.

[16] From Chakravartty and Anjan, "Scientific Realism," *The Stanford Encyclopedia of Philosophy* (Summer 2011 Edition), E. N. Zalta (ed.), http://plato.stanford.edu/archives/sum2011/entries/scientific-realism/. I consider only the physical world, not social or political "worlds" for the purposes of this work.

[17] This antirealist doctrine was primarily explicated in Berkeley's *Treatise Concerning the Principles of Human Knowledge* (1710).

(discussed previously in Chapter 2) is relevant to the realism/antirealism dichotomy because Kant asserted that the only world we can ever come to know is that which depends on the concepts and functions of the human mind: the world of appearance, or what he termed the "phenomenal" realm. Kant did assert that there was "something else out there"; in his terms, the "noumenal" realm, but it was a basic principle of his philosophy that we can never come to know this elusive realm, that which he called the "thing-in-itself." Devitt (1991) refers to Kant as a "weak realist" because Kant did hold that there was *something* that existed independently of our knowledge, even if we could (according to Kant) never obtain knowledge about it.

In the latter twentieth century, Kant's basic approach evolved into a version of antirealism generally known as "constructivism." In Devitt's terms, constructivism asserts that "we make the known world" (Devitt, 1991, p. 236). He correctly (in my view) points out that much of the constructivist argument rests on a conflation of epistemological (knowledge-based), semantic (meaning-based), and ontological (metaphysical) issues. But despite these weaknesses in the usual sorts of arguments for constructivism, it is in quantum theory where this form of antirealism begins to gain traction because of the notorious dependence of property detection on what we choose to measure (recall Section 1.1). In contrast, realism demands that the object of knowledge is *not* fundamentally changed by observation.[18]

We can formulate this dispute in terms of the *subject–object distinction* presupposed by any discussion about knowledge on the part of an observer (subject) and the aspect of the world he wishes to know about (object). In these terms, the realist believes that knowledge is *object-driven*, while the antirealist believes that knowledge is *subject-driven*. We can now make contact with PTI by identifying the "object" with the offer wave and the "subject" with the set of confirmations taking place upon absorption of the offer wave components. The latter can be thought of in terms of a particular experimental setup or just in terms of the sense organs of an observer.

With the above identification, PTI can resolve the realism/antirealism conflict by declaring a measured form of "victory" for both sides. Realism correctly asserts that there truly is "something out there" that is independent of observation. In PTI terms, this is the object represented by a quantum state or offer wave $|\Psi\rangle$. But antirealism correctly asserts that the form the "something" takes is at least partly dependent on

[18] The Bohmian theory provides a way to retain realism about quantum objects because it asserts that there really are quantum particles with definite positions, independently of our knowledge or concepts. (Bohmians acknowledge that we disturb those positions in an uncontrollable way when we measure certain contrasting (non-commuting) properties, but that if we choose to measure position, what we find is a particle position that existed independent of our observation. However, I do not favor the Bohmian theory because the "guided particle" ontology is incompatible with the relativistic domain (e.g., recall that the classical electromagnetic field must be described by an indefinite number of quanta); there is no account of how guiding waves living in $3N$-dimensional configuration space "guide" particles in 3-space, and its account of the Born Rule can be only statistical in nature.

Figure 7.5 Subject and object.

how it is observed (in physical terms, detected in an actualized transaction), which takes into account the types of confirmations $\langle\Phi|$ generated by absorbers. Recall from Chapter 4 the man observing the table, reproduced here as Figure 7.5. It's not the "categories" or "concepts" in his mind that do the primary work here, but simply the absorbers in his sense organs. Thus, the "subject–object" dichotomy becomes the "confirmation–offer" complementary relationship in PTI.

The foregoing "defangs" antirealism in the following sense: it need not be anthropocentric, since in PTI, one can have an actual phenomenon/event in the absence of a "conscious observer." All one needs is emitters and absorbers, which are physical entities.

This formulation also provides a solution to a long-standing puzzle faced by Kant scholars. The problem is this: Kant insisted that knowledge of the phenomenal world was obtained by way of an interaction of human perceptual activity and concepts with the noumenal world. But the nature of this interaction was deeply obscure. If the noumenal object or "thing-in-itself" was truly "unknowable," what sort of causal power could it have to produce knowledge, even if through human-centered concepts and perceptions? PTI provides at least a partial answer: the noumenal realm is the realm of offer waves; the phenomenal realm begins with their absorptions, which generate confirmations and ultimately specific actualized events. The nature of the interaction between the noumenal realm and the phenomenal realm is just the transactional process.

Thus, in Kantian terms, one can say that the knowable phenomenon is rooted in the unknowable noumenon (quantum entity or offer wave) which is answered by confirmations from absorbers in the sensory organs. Actualized transactions result in transfers of energy, which are processed by the senses and their attendant cognitive structures. There are two components to the latter process: (1) physical/ontological (the quantum transaction arising from absorption by the sense organs) and (2) epistemic (the subjective/theoretical concepts used to identify and understand the phenomenon arising from the transaction). The current work deals only with aspect (1) because that is all that is necessary to account for the basic phenomena (the "raw sense data" as described in a Russellian or foundationalist account).[19] As has been noted by other researchers (e.g., Kent, 2010), having to bring in philosophies of mind or explicit psycho-physical dualism weakens the scientific account because there is no account of "mental substance" in the exact science of physics. Traditional "collapse" approaches inevitably must engage in forays into psychologism of this kind because there is no consistent way to break the linearity of the theory and thereby provide for a determinate result on the physical level without taking absorption into account.

Thus, the transactional model denies the strongest form of realism, namely the view that objects in their independent entirety are "directly given" to the senses; but it provides support for what is termed "representational realism." The latter assumes that what is directly present to the knower is not the object itself, but "sense data" that make contact with the objectively existing external object and therefore provide authentic knowledge about it. In PTI, sense data are the product of the object, as a source of offer waves – and the subject, as a set of absorbers. Together, the subject and object produce transactions that provide information about the object *conditioned on the manner and circumstances under which it is perceived.* The latter sentence is important: such knowledge is always only partial, since transactions vary depending on what types of absorbers are available to the offer waves comprising the object.

7.5 An example: phenomenon vs. noumenon

This section makes contact with Shakespeare's famous verse that opened this chapter. Let us consider an example of the way in which a *phenomenal* world of appearance, thought of as occurring in "spacetime," arises from a transcendent *noumenal* level in terms of an aspect of popular culture: Internet-based "massive multiplayer online role playing games" or MMORPGs, such as "World of Warcraft" or "Second Life."

In the game Second Life, a player can access an online game environment by loading a software package on his local computer. The player uses the software to create for himself a character, or "avatar," which represents him in the online game environment. Let's call the human player "Jonathan" and his game avatar "Jon."

[19] In this regard, I do not deal in this work with the deep and subtle questions concerning the relationship of subjective perception to sense data, although I do assert that perception properly needs an object, even if not "physical" in the usual sense: perception is transitive and presupposes the fundamental subject–object distinction. (In contrast, one might refer to a perception-free account of experience as *awareness*, which is the ability to perceive.) I assume that *whatever* it is that is subjectively perceived can be attributed to physical transfers of energy via actualized transactions. In cases of non-veridical or hallucinatory perception, an account may be possible in terms of atypical biological processes in the hallucinating subject which ultimately can be traced to transactions among the microscopic constituents of biological components (e.g., neurons).

Once Jon is established in a game environment, he carries with him a point of view (POV) through which Jonathan can perceive what Jon perceives as the latter pursues his in-game career. Now, suppose Jonathan decides to have Jon create something – a table, for example. Jonathan can input certain commands through Jon into the game environment, and a "table" will appear at the desired "location" in Jon's vicinity.

Now, consider another human player, Maria, whose game avatar is "Mia." Maria might be sitting at her computer in Sydney, Australia while Jonathan is in Montreal, Canada. Nevertheless, their avatars may be in the same game environment "room," say the "Philosophy Library," where Jonathan/Jon has just created his "table." Now, suppose Jon and Mia don't know that they are only avatars, but assume themselves to be autonomous beings. We might imagine Jon and Mia discussing the table in front of them along the same lines as the discussion in Bertrand Russell's *The Problems of Philosophy*, Chapter 1. For readers unfamiliar with this material, Russell's discussion involves noting that the appearance of the table depends, to a great extent, on the different conditions under which it is viewed (or, more generally, perceived). These appearances may be mutually contradictory: for example, the table may appear smooth and shiny to the eye, but rough and textured under a microscope. Following this line of argument, Russell famously concludes that the only knowledge we can have of the table is of various aspects of its *appearance*, which must always be contingent on the conditions under which it is perceived; and that the "real" table underneath the appearances – whatever that might be – is a deeply mysterious object. In his words: "Thus it becomes evident that the real table, if there is one, is not the same as what we immediately experience by sight or touch or hearing. The real table, if there is one, is not immediately known to us at all, but must be an inference from what is immediately known. Hence, two very difficult questions at once arise; namely, (1) Is there a real table at all? (2) If so, what sort of object can it be?" (Russell, 1959, p. 11). Russell's presentation is an account of the deep divide between, in Kant's terms, the world of appearance (phenomenon) and the thing-in-itself (noumenon). (Notice how he repeats the phrase "if there is one," to emphasize how little we really know about it.)

If Jon and Mia pursue this analysis they, too, find that the only knowledge they have of the table is based on its appearance (which their human players can monitor on their computer screens showing their avatars' POVs). Suppose the side of the table first facing Jon is black and the other side, facing Mia, is white. Jon and Mia can talk to each other and discuss what they see, and they can agree to compare their perceptions by, say, changing places. Then Mia can confirm that the other side of the table is black, and vice versa. By performing this sort of comparative observations, Mia and Jon can convince themselves that there "really is" a table there because they can *corroborate* their different perceptions in a consistent way: their intersubjective observations form a coherent set. This suggests to them that there is "something out there" that is the direct cause of their perceptions. In commonsense realist fashion, they might conclude that there is a "real" table behind or underneath the appearances – a "table-in-itself" – that "causes and resembles" their perceptions of it.[20]

But what about Jonathan and Maria? They both know that, while the "table-in-itself" could be said to be the cause of Jon and Mia's perceptions of the game table, the "table-in-itself" *does not "resemble" the game table at all.* What is the "table-in-itself"? It is nothing more than *information* in the form of binary data, manipulated by the people who created the game and by the human users (Jonathan and Maria). Compared to the game table perceived by Jon and Mia, it is insubstantial, abstract. And yet clearly, it *is the direct cause of the avatars' perceptions of an ordinary table* (the "table-of-appearance") which, to them, is *not* just an "illusion": the avatars cannot ignore it (for example, they will bump into it and may even incur physical damage if they try to run through it as if it isn't really there). If a human user were to somehow speak to an avatar like Mia and tell her that the objects in her world are nothing but information, she would scoff at the suggestion, and might ask why she suffers damage if she falls off a cliff in her "only information" world. To the avatars, their world is perfectly concrete and consequential.

What does this little parable tell us about our world of "ordinary" objects-of-appearance; that is, our empirical world? It tells us that it is conceivable and even quite possible that the "table-in-itself" of *our* world is a very different entity from what the table-of-appearance might suggest. Because we, and the objects around us, are governed by the laws of physics (the "rules of the game," if you will), we interact with them and are affected by them, and in that sense they are certainly real, just as the game-environment objects are real for Jon and Mia. But the "object-in-itself" is precisely *that aspect of the real object which is not perceived.* If such an aspect exists at all, we can reasonably expect it to be on an entirely different level from our perceived world of experience. Indeed, in terms of PTI, the "object-in-itself" can be considered to be the offer wave(s) giving rise to possible transactions establishing the appearances of the object. Just as the "table-in-itself" behind the avatars' table does not really live in their game world and is a kind of abstract information, so the offer waves giving rise to our real empirical objects do not live in spacetime and can be considered a kind of abstract but physically potent information – i.e., the physical possibilities first introduced in Chapter 4.

Now, recall from Chapter 2 that Kant asserted that the "thing-in-itself" is *unknowable.* I wish to contest this, based on two main (disparate) points: (1) the fact that Kant has already been shown to have been mistaken in assuming that Euclidean

[20] The naïve realist notion that independently existing objects outside the mind are the causes of ideas (perceptions) that resemble them is extensively critiqued in Descartes' *Meditations*.

(flat) space is one of the "categories of experience"[21]; and (2) the fact that *perceiving* (i.e., sensory perception) is not equivalent to *knowing*, since knowledge can also be obtained by intellectual (rational) means.[22] Concerning (2), recall the arguments in Chapter 2 that an empirically successful principle-type theory can be taken as providing new theoretical referents to previously unknown structural properties of the world. Such an approach to new knowledge is an *intellectual* or rational one rather than an empirical one, the latter being dependent on observation through sensory perception (including the use of sense-enhancing technologies such as microscopes or telescopes), and therefore being subject to the limitations of appearance. In contrast, unexpected but fruitful theoretical development can be considered as pointing to an abstract (non-observable) level of reality inaccessible to observation, as in the postulation of atoms. The latter was an intellectual step forward in knowledge, not an empirical one.

Recall also that Bohr asserted that the quantum object is something "transcending the frame of space and time" – suggesting (albeit despite himself) an altogether metaphysically new type of entity. The Hilbert space structure of quantum theory greatly exceeds the structure of the empirical world in that it precludes our ability to attribute always-determinate classical properties to objects (recall Chapter 1). Therefore, it's natural to suppose that the structure of the theory describes something "transcending the frame of space and time" but which is nevertheless real because objects described by those Hilbert space states can be created and manipulated in the laboratory.[23]

Let us review the argument so far: players in an online game such as Second Life (SL) can intersubjectively confirm the existence of an object in the SL environment, just as people in our world can intersubjectively confirm the existence of a table. But the object-in-itself remains elusive, in that each observer who perceives the object perceives a different version of it. That's because the object-in-itself exists in domain II (recall Figure 7.2); it is not observable because it is not actualized and therefore does not exist in the world of appearance (i.e., "spacetime"). At the game level, the object's observation by a particular avatar Mia is contingent on a transaction between the avatar and an aspect of the object, that aspect being determined by the manner and circumstances under which the object-in-itself (OW) is received – i.e., the confirmation wave generated by the avatar. The CW consists in the user

[21] This could be considered the "Kant's credibility is already suspect" argument.

[22] That this is the case is demonstrated by the great empirical success of physical theories arrived at through rational analysis and mathematical invention. In Einstein's words: "How can it be that mathematics, being after all a product of human thought independent of experience, is so admirably adapted to the objects of reality?" (2010). Nature seems to be inherently mathematical and logical; were that not the case, theoretical science could not provide any useful knowledge.

[23] Here I endorse Hacking's dictum that "if you can spray them then they are real" (Hacking, 1983, p. 23), referring to an experimentalist's comment that he could "spray" a piece of equipment with positrons.

Maria turning on her computer, loading the game, accessing a particular location in the game world, and orienting her avatar Mia's POV in the appropriate direction (all these being dictated by the *information* of domain II which is the data manipulated by the human players Jonathan and Maria). The "actualized transaction" consists in Mia's POV registering the appearance of the table by specifying which pixels on the screen should be colored red, green, etc. This is only possible because of two things: (1) Jon/Jonathan created the table "offer wave" with specific properties and (2) Mia/Maria accessed the appropriate properties in order to receive the "offer wave" and actualize its appearance in her POV.[24]

We can use this model to immediately gain insight into the phenomenon of "non-locality." While the avatars and their objects have a maximum speed *c*, Jonathan and Maria transcend the game environment and can freely communicate instantaneously (with respect to the game environment), so that information can be transmitted from one region in the game environment to any other at infinite speed. This is precisely because that information is *not actually contained in* the game environment. So, for example, Mia might shoot an arrow at game-speed *c* in Jon's direction while Maria tells Jonathan (over the phone) that she is doing so. Instantly, Jon can step aside and miss the arrow, even though he should not be able to do so according to the rules of the game environment (which would preclude Jon from seeing the arrow coming at him). "Faster-than-light" or "non-local" influences are evidence of physically efficacious information existing on a level other than that of the usual local processes (i.e., the game environment or "spacetime").

7.6 Causality

In this section, I consider the vexed notion of "causality" and discuss how transactions can illuminate this longstanding conundrum.

7.6.1 Hume's elimination of causality

The reader may recall that the Scottish philosopher David Hume first cast enormous doubt on this commonplace notion of everyday life. As a strict empiricist, he looked for specific evidence of causality in the empirical (observable) world and could not find it. For example, consider a billiard game. The player strikes the cue ball; the cue ball moves and strikes another stationary ball. Subsequently, the second ball moves with the same momentum as the cue ball, which comes to a halt. It is perfect common sense that the cue ball *caused* the second ball to begin

[24] At a higher meta-level are the game designers who decide what types of OW can be created and how – the "Gods of the Game," if you will.

moving. However, we never actually *see* the cause; all we see is the pattern of events, which is repeated every time we perform these actions. The reader may object: but surely, we saw the cue ball strike the second ball. How could the second ball *not* move, since it was hit by the cue ball, which we clearly observed? But notice again that we did not actually see the cause; the cue ball striking the second ball is not *observably* a "cause." It is simply an event. Our expectation that the second ball must move is based on the fact that we have always seen this happen. It is certainly conceivable that the second ball could just sit there, despite having been hit. The motion of the second ball is predicted by physical law; but again, physical law simply describes patterns of events; it does not say *why* they happen. For this reason, Hume concluded that causation is not really in the world, but is something we *infer* from what he termed the "constant conjunction of events."

Another aspect of the "common sense" of causality (despite the fact that we never actually see it) is that the cause always precedes the effect: in terms of the above example, the cue ball striking the second ball *precedes* the motion of the second ball. The contingent, empirical time-asymmetry of causation is addressed further in Chapter 8. For now, I note that this feature of causation is simply a feature of the types of patterns that we see in the empirical world, and should not be thought of as necessarily extendable to the unobservable entities of the micro-world (e.g., electrons), as is customarily assumed.

7.6.2 Russell, Salmon, et al.

As might be expected due to its unobservable nature, the concept of causality is a very slippery and elusive notion. Many distinguished philosophers have attempted to chase it down and capture it in definitive terms, without conclusive success. Bertrand Russell initially expressed great skepticism about causality in this famous quip:

> The law of causality, I believe, like much that passes muster among philosophers, is a relic of a bygone age, surviving, like the monarchy, only because it is erroneously supposed to do no harm. (1913, p. 1)

Russell nevertheless felt that causality needed to be well-defined in order to support the development of physical laws which seemed to imply causal processes (even if physical laws do not explain them). He developed a theory of causality in terms of "causal lines" (Russell, 1948). This theory was based on several reasonable postulates, such as the idea that there is a kind of "quasi-permanence" in the world: we do not see utter chaos, with objects suddenly and randomly changing their properties. However, Russell's theory was far from bullet-proof, and came under sustained and

Figure 7.6 The moving spot that "exceeds the speed of light."

cogent criticism from Wesley Salmon (1984), who proposed his own theory of causation. Salmon sought to distinguish genuine causal processes from "pseudo-processes" consisting of effects which are not causal in the usual sense. An example is a moving spot of light on a wall which can exceed the speed of light (see Figure 7.6). In that case, no material object actually exceeds the speed of light, but an observable artifact does.

Salmon endeavored to capture the essence of causality in terms of the ability of a causing event to transfer a "mark" to the affected event (some persistent change in the second event which is the effect). However, this theory, too, has been found to have loopholes that hinder its ability to distinguish between what we consider to be genuine causal process and pseudo-processes, such as the moving spot of light or the changing portions of a charged metal plate in shadow (Salmon, 1997, p. 472). Another weakness, according to critics, is that it does not take into account processes such as trajectories of bodies, in which an earlier state seems to serve as the "cause" of a later state. This issue is related to the notorious philosophical riddle of identity and persistence of particular objects. The story of the philosophical pursuit of "causality" as an ontological entity is thus one of the attempts to construct theories of causality which exclude all situations that we regard as non-causal and include only those that we regard as causal.

There has been no conclusive resolution to this puzzle, and I suggest that this is because Hume and Russell were right: causality is *not* an ontological feature of the world. In TI terms, it is an inference we make based on situations involving very probable transactions (i.e., transactions with weight close to unity). It can be seen as a supporting feature of physical law because overwhelmingly probable transactions underlie the empirical expression of such fundamental physical principles as the "principle of least action" (recall Section 4.4.1).

7.6.3 Transactions to the rescue

Note that we can also understand the distinction between "causal processes" and pseudo-processes in terms of the transactional process. A transaction constitutes a transfer of energy from the emitter to the absorber. The spot in Figure 7.5 is the location of an emitter (it is a point of reflection of photons, and in microscopic terms[25] this means that a photon offer wave is annihilated and a new one emitted at that point). Thus the *location* of photon annihilated emission is moving at a speed greater than c, but no energy is actually being transferred faster than c. We can also account for the apparent persistence of macroscopic physical objects in terms of transactions; recall the baseball of Section 7.4.1, whose apparent persistence depends on transfers of energy via transactions among its constituents and between those and our sense organs. If "earlier states cause later ones," it is in terms of such energy transfers.

Other pseudo-causal processes can similarly be ruled out by reference to transactions. For example, transactions allow us to unambiguously demarcate genuine persistent objects from pseudo-persistent "non-objects," such as the parts of a charged metal plate in shadow, only when they are in shadow (Salmon, 1997, p. 472). Dowe's reply (2000, pp. 98–9) is that this is not a causal process because the above is not a genuine object – it does not possess identity over time. The burden is then on Dowe to define what constitutes identity over time – which he takes as primitive and thereby, according to Psillos (2003, p. 124), makes his account circular. We can define the persistence of an object through time as attributable to ongoing transactions among its constituents as discussed above in connection with the baseball example. The charged metal plate is a network of transactions whose macroscopic cohesiveness is supported by those transactions; the changing set of portions of the charged plate in shadow is not. The latter consists of the changing set of transactions between an observer and emitter(s) outside the plate; i.e., between a light source, an observer, and some object making certain portions of the plate inaccessible to the light source (hence resulting in the appearance, to the observer, of shadowed regions where the emitted OW cannot be reflected from the plate).

7.7 Concerns about structural realism

I conclude this chapter by considering a higher-level issue of interpretive methodology. I noted earlier that PTI can be considered in a weaker, structural realist (SR) form which remains agnostic about what these sub-empirical offer and confirmation waves "really are" in ontological terms. In that regard, I should address some

[25] That is, in terms of quantum electrodynamics in which reflection is a type of scattering event: the incoming photon OW of momentum p is distinct from the outgoing photon OW of momentum p'. See also Feynman (1985, p. 101).

objections to SR, which was first developed by Worrall (1989) in an attempt to circumvent the so-called "pessimistic induction" concerning the ability of scientific theories to refer to ontological entities. The "pessimistic induction" consists in pointing out that many of those supposed entities (e.g., "phlogiston") were later found not to exist; thus, based on past experience, it seems likely that the putative entities referred to by a currently successful theory might also be repudiated. Worrall proposed instead that successful theories refer to *structural* aspects of the world, even if it could not be known what the specific nature of those structures were.

Psillos (1999) has objected that Worrall's distinction between structure and nature (i.e., substance/properties) cannot be maintained when applied to specific entities described by a theory:

> To say what an entity is is to show how this entity is structured: what are its properties, in what relations it stands to other objects, etc. An exhaustive specification of this set of properties and relations leaves nothing left out. Any talk of something else remaining uncaptured when this specification is made is, I think, obscure. I conclude, then, that the "nature" of an entity forms a continuum with its "structure", and that knowing the one involves and entails knowing the other. (pp. 156–7)

The above characterization applies to empirical phenomena, perhaps, but not necessarily to sub-empirical entities. That is, one can consistently propose that the structure of quantum theory dictates that the entities described by the theory cannot be considered to exist within the confines of a spacetime manifold (since the relevant mathematical space for N quanta is $3N$-dimensional and therefore not mathematically commensurate with spacetime). Therefore, we can remain agnostic about the precise nature of those entities but still insist, based on empirical success of the theory, that their dynamical *structure* is accurately captured by the form of the theory. The theory says how the entities are structured but not what they are: in Aristotelian terms, it provides their "formal cause" but not their "material cause" (if any!).[26]

Thus the key difference between the current proposal and typical structural realist proposals is that it denies the usual unexamined identification of "real" with "empirical." For example, Barnum (1990) offers the following comment concerning a formulation by Dieks:

> In Dieks' view, his semantical rule is the sort of thing which is necessary in any attempt to interpret a physical theory: "certain parts of the models [of the theories] are to be identified as empirical substructures; i.e., part of the theoretical models have to correspond to observable phenomena." I agree with this general characterization of the interpretation of theories: the "internal meaning" of the terms of the theory, given by the mathematical

[26] The ancient Greek philosopher Aristotle proposed that all objects have four types of cause: material (relating to its substance); formal (relating to its structure); efficient (relating to its creator); and final (relating to its purpose).

structures which are models of the theory, needs to be supplemented by "empirical meaning." This is done by showing how the theory relates to our experience. (p. 2)

This characterization certainly has had its merits in connection with classical theory, in which all physical entities can be considered as existing in spacetime. However, the above approach would seem too restrictive for quantum theory, whose structure is incommensurable with the empirical arena of spacetime. We already know what parts of quantum theory relate to our experience – i.e., the probabilities given by the Born Rule – but the point of a realist interpretation of the theory is to go *beyond* that, to find a physical referent for those parts of the theoretical model that cannot be identified as empirical substructures. Thus I agree with Ernan McMullin (1984), who notes that part of the interpretational task is to discover to what the theoretical quantities refer, without assuming that they must refer to something in the macroscopic (empirical) world:

[I]maginability must not be made the test for ontology. The realist claim is that the scientist is discovering the structures of the world; it is not required in addition that these structures be imaginable in the categories of the macroworld. (p. 14)

McMullin's point above is a subtle but crucial one, which cannot be overemphasized in connection with the present work. Specifically my claim is that quantum states refer to something sub-empirical, yet real. As noted previously, this is a new category which is not part of the macro-world, and it is not legitimate to reject it based merely on perceptions that it might seem "implausible" or "unimaginable" when compared to the categories of the macro-world. So it is bound to be counterintuitive. Yet one can still show "how the theory relates to our experience" by positing the conditions (i.e., the actualizing of transactions) under which the sub-empirical entities give rise to empirical events.

Psillos' objection thus begins with a premise with which I would disagree; namely, "To say what an entity is . . .": a structural realist is not committed to the claim that a theory always says what an entity *is* – that it gives an "exhaustive specification" in usual spacetime or substance/property terms. In fact, this was exactly Newton's interpretive stance when asked to what "gravity" refers.[27] Newton clearly regarded his theory as *about* gravity and as *referring to* gravity; thus he was realist about his theory. But his theory did not spell out the specific ontological nature of gravity.[28]

[27] Concerning the ontology of gravity, Newton famously stated "*Hypotheses non fingo*" (I feign no hypotheses); from his *General Scholium* appended to the *Principia* of (1713).

[28] A similar argument is presented in Dorato and Felline (2011): ". . . we propose, therefore, the properties of the *explanandum* are constrained by the general properties of the Hilbert model [of quantum theory]. In this sense the *explanandum* [e.g., how or why quantum systems obey Heisenberg's uncertainty principle] is made intelligible *via* its *structural similarities* with its formal representative, the *explanans* [e.g., representability of

PTI in its strong form does go beyond the original TI by proposing a specific ontological referent in the form of physical possibilities. Nevertheless, if one is reluctant to embrace this new metaphysical category, one can still allow that TI captures an essential *structural* element of quantum systems (advanced solutions arising from absorption) missing in the usual account, and thereby provides a more complete interpretation than its competitors.

In the next chapter, I consider the nature of spacetime in PTI.

such systems by Fourier expansions]. Given the typical axioms of quantum mechanics . . . any quantum system exemplifies, or is an instance of, the formal structure of the Hilbert space of square summable functions." (2010, p. 6; preprint version)

8

PTI and "spacetime"

8.1 Recalling Plato's distinction

Let us begin this penultimate chapter by recalling the philosophy of Plato, discussed briefly in Chapter 1. Plato distinguished two levels of reality: (1) "appearance" and (2) "reality." In Plato's philosophy, (1) means the world as directly perceived by the five senses, and (2) means the unperceived world that is understood by the intellect. In modern terms, these two realms would be called (1) the empirical and (2) the ontological (or extra-empirical) realms, respectively.

Now, the traditional task of physics is to attempt to describe *all* of reality – including that which is not apparent – by accurately observing and insightfully analyzing the world of appearance using logic and mathematics. In other words, physics studies the empirical realm in order to understand *both* the empirical and extra-empirical realms. (A strict empiricist would deny that the job of physics is to gain knowledge of an extra-empirical realm even if it exists. But that approach can be seen as evasion of the scientific mission, as argued in Chapter 2.)

8.1.1 What is the empirical realm?

First, let us consider the question: what exactly *is* the empirical realm in strict physical terms? It is often thought of as all of spacetime. However, this cannot be right if the empirical realm is precisely the world of appearance – of direct experience – since we can experience neither the past nor the future (even with powerful instruments).[1] All we can experience is the present, the "now" as it is presented to our senses. So, if we really want to be careful about it, *only the now is*

[1] This holds for observations of distant astronomical objects, as follows. The light we detect from a galaxy 10 thousand light years away left that galaxy 10 thousand years ago, but we don't actually see it until it reaches us in the present. Thus, we see the galaxy in the present as it was 10 thousand years ago. We don't actually experience the past. This is the same as getting a message in a bottle from a castaway. The castaway may be long dead, but the message is something he wrote while alive.

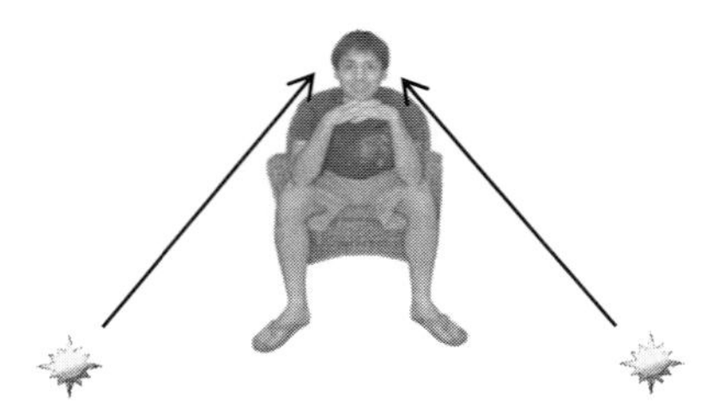

Figure 8.1 The "now" is the empirical realm.

the empirical realm. What do we directly experience about now? That it presents properties to us which are always changing.[2] That is, *the now does not "move"; it changes*. How do we experience these changing properties? We experience them by way of electromagnetic signals that transfer energy from what we are observing to our sense organs (by way of actualized transactions). Thus, our "now" is defined by a spatial coordinate (or, in a relational view of spacetime,[3] the object(s) with which we are currently in direct contact) and any light signals that have reached our eyes from other objects. This is illustrated in Figure 8.1, where the now is symbolized by the person's chair (let's call him "Ty") and the light signals reaching his eyes from objects in his past.

The foregoing raises a further issue concerning the objectivity of the empirical realm: it *cannot be "objective" in absolute terms because it is defined in terms of appearance, and appearance can only be relative to a given observer.* This means that, strictly speaking, everyone has a different "empirical realm." However, we can corroborate our experiences and arrive at a consistent *intersubjective* consensus about a "larger" world of appearance beyond our individual empirical realms. All of these corroborations are conducted using electromagnetic waves. Thus, what is referred to as "the empirical realm" in physics can be no more than that corroborated collection of individual empirical "nows."

The above reinforces the idea that relativity theory (with its limitation of signal speeds to the speed of light) places restrictions on the empirical realm. But the empirical realm is in fact even more restricted than is often noticed in discussions of relativity. For example, consider a typical spacetime diagram such as Figure 8.2.

Besides the diagonal "light cone" lines, the diagram indicates gray horizontal "lines of simultaneity" for a given reference frame. But even though these lines are defined with reference to a given observer (Ty), almost all of the points comprising those lines are extra-empirical: they are not within Ty's empirical realm. For

[2] Norton (2010) makes this point as well: ". . . we do have a direct perception of the changing of the present moment. That is clearest in our perception of motion."

[3] I discuss relationalism below. In a nutshell, relationalism denies that spacetime exists as an independent substance or "container."

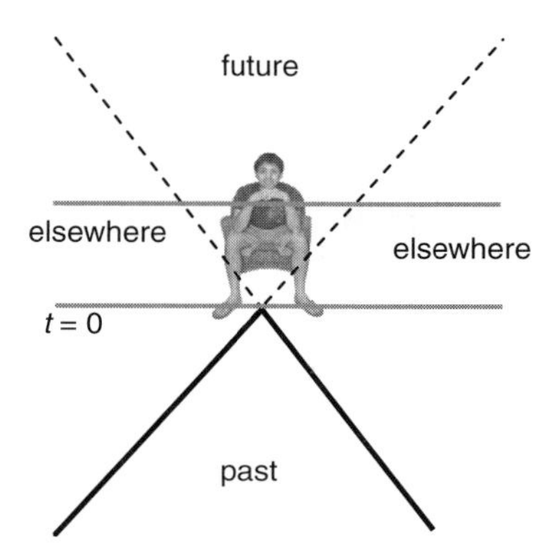

Figure 8.2 A spacetime diagram.

example, the line $t=0$ crossing Ty's "now" is extra-empirical (note that it is in the "elsewhere" region) except for those points in direct contact with him. In three dimensions, these are hyperplanes of simultaneity: spheres with respect to the three spatial dimensions, each with a particular time index. Owing to the fact that electromagnetic signals have a finite speed, no observer *really* sees or touches anything outside the apex of his light cone.[4] When you sit in a chair reading this book, you are seeing the page as it existed a few nanoseconds ago, *not* as it exists along a line of simultaneity from your eyes.

8.1.2 The past vs. the future

Note in Figure 8.2 that the future portions of Ty's light cone are dashed, while the past portions are solid. This is meant to indicate that in PTI, the future is not actualized but exists only as possibilities.[5] On the other hand, many features of the past have been actualized; but not necessarily all, as discussed in Chapter 5 in connection with contingent absorber and delayed choice experiments. In those experiments, and indeed in any situation in which there is uncertainty about where or when an offer wave will be absorbed, the past remains indeterminate until the absorption is specified. (The previous sentence implies what philosophers term an "A-series" view of time, which will be discussed in more detail in the next subsection.) Nevertheless, the fact that events can be actualized and also be in the

[4] It may seem lonely to realize that each of us sits isolated atop the apex of our light cone. But recall Black Elk's vision: "'I saw myself on the central mountain of the world, the highest place, and I had a vision because I was seeing in the sacred manner of the world.' ... And then he says, 'But the central mountain is everywhere'" (Neihardt, 1972). This can just as easily describe the way each observer is at the top of his own empirical "mountain."

[5] This basic picture of time is termed "possibilism" in Savitt (2008), although the present model differs from the usual "growing universe" or "possibilist" temporal theory, as will become apparent.

past means that *events may be actualized but not contained within a given observer's empirical realm or "now."*

The idea that an *actual* event may not be *empirical* for a particular observer may seem strange at first, since the two adjectives are usually taken as synonymous. But this is necessary, since no two observers really share exactly the same "now," and any event experienced by any observer has to be an actualized event. But recall also that under PTI, any event corresponding to the actualization of a transaction is an actual event, whether or not an "observer" is present to witness or record that event. So actualization of an event is a necessary but not sufficient condition for that event to be empirical relative to any particular observer.

In addition, the assertion that there are no actualized events in the future light cone of any observer marks a significant divergence between the spacetime ontology of PTI and the usual "block world" view, so let's dwell on that for a moment. What it means is that, while we could certainly draw another light cone centered on a hypothetical observer in the future light cone of Ty, that image would be merely a conceivable possibility (category III in Figure 7.2) which does not correspond to our world.[6] This is because all physical entities were at the same point in the primal "now" (i.e., the Big Bang), and events are actualized in the now and recede from the now into the past (see Section 8.1.3 below for elaboration). Since, according to the relational spacetime ontology proposed here, the "fabric of spacetime" is no more than the structured set of actualized events in the past or elsewhere, there can be nothing actualized in any observer's future light cone – including, of course, another hypothetical observer.[7]

Thus the spacetime diagram, because it is so easily subject to arbitrary event placements in a hypothetical "future," typically misleads us into thinking that there can be "future events" and "future observers" when – according to the model proposed herein – this is physically not the case. Just because we can draw something on a spacetime diagram does not mean that it can physically exist in our world. The notion that the ability to represent something on a spacetime diagram implies that it may physically exist can be very compelling.[8] However, to see why we need to be wary of subscribing to this unwarranted assumption, consider the following analogy. Animation artists now have programs that can do a lot of the tedious work of redrawing frame after frame of the same character for them. A typical animation program allows you to load a basic image of a character, indicating where all the joints are, and the program will change the angles of the joints for you in a series of images to make the character appear to move. You have to specify the amounts by which each of the joints is to move in each

[6] This is the usage of "possibility" corresponding to the non-physical, unreal possibilities of category III in Chapter 7.

[7] A possible objection to this account is addressed in Section 8.1.4 below.

[8] This is basically the same point, albeit with respect to "spacetime" rather than just time, made in Norton (2010) in a slightly different context: "We start to get used to the idea that our theories of space and time are telling us all that can be said about time objectively" (2010, p. 26).

frame. Theoretically (for example), you could make a character's head turn by any amount in any direction, but that doesn't mean that the motion will be realistic or even physically possible. The spacetime diagram similarly lacks certain physically relevant constraints. So the freedom to draw whatever we choose, wherever we choose on such a diagram does not imply that what we drew corresponds to what is physically possible, any more than the freedom to make a character's head spin around in circles in an animation program means that this would be possible in the real world. As observed in Chapter 2, the map (i.e., the spacetime diagram) is *not* the territory, and can correctly represent only certain specific aspects of the territory.

8.1.3 The fabric of created events

To gain further insight into the proposed spacetime ontology, consider the following metaphor. Think of the past as a knitted fabric (see Figure 8.3). The "now" is the set of stitches on the knitting needles, whatever the time index t (here, indexing the row) of those stitches. Let's assume that $t = 0$ corresponds to the Big Bang, when the first stitch is "cast on" to the needle. The future is nothing more than one or more balls of yarn of different types, a pattern, and/or some ideas about what to knit. The "now" is the realm in which our garment is created; the now doesn't "move," but the

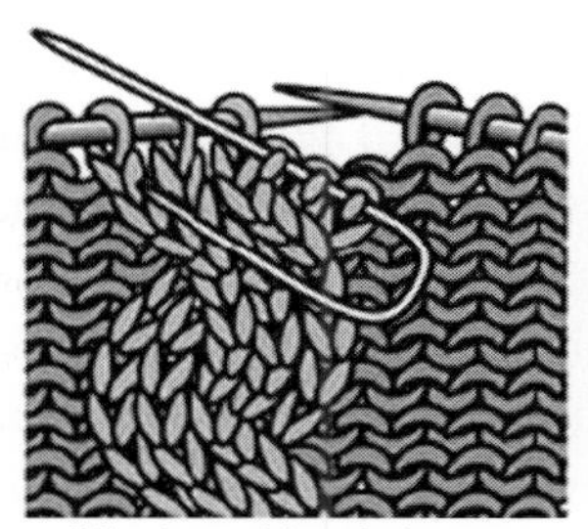

a. Slip the cable stitches to the cable needle and hold in front.

b. Knit 3 from the LH needle.

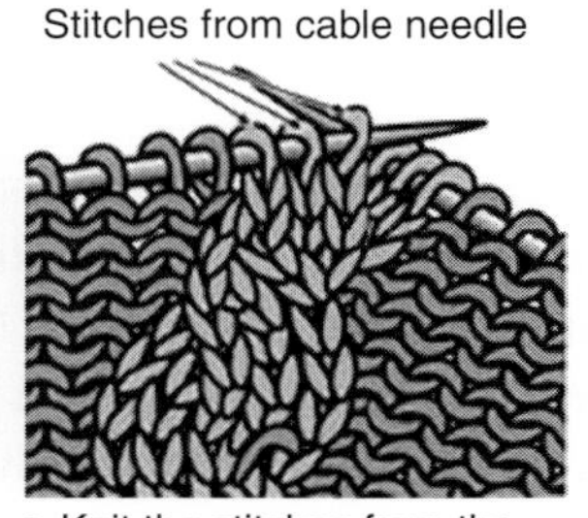

c. Knit the stitches from the cable needle.

Figure 8.3 The past as a knitted fabric; "now" corresponds to the set of stitches on the needles. This picture is a version of C. D. Broad's "growing universe" theory of time (Broad, 1923). It has much in common with the theory of Tooley (1997) in that a statement like "the stitches in row t are currently on the needle" can be seen as equivalent to "event E is present at time t" in Tooley's picture. This is because, for the former statement to be true when uttered, there is not yet a row $t+1$ (i.e., the future is not actual). However, I obviously do not adopt other features of Tooley's theory, such as its spacetime substantivalism.[9]

[9] Image from http://www.dummies.com/how-to/content/knitting-cables.html. From *Knitting For Dummies*, 2nd edn by Pam Allen, Tracy Barr, Shannon Okey. © 2008. Reprinted with permission of John Wiley & Sons, Inc.

stitches on the needles change (perhaps in color or texture) and are extruded away from us in the form of fabric as the knitting progresses. The creation of a new "row" is always attributable to the actualization of specific transactions, as discussed above. Thus, the now is not something that "moves forward"; rather, the now is the empirically always-present field of change, while *the past is something that continually falls away from us.*[10] So, a locution such as "when $t=5$ is now" means the stage of the knitting process at which the stitches time-indexed by "row" number 5 are on the needle (in more familiar terms, "are in the present"). As the process continues, those stitches are extruded and are no longer on the needle, but they keep their time index as they recede.[11] Thus, if we are knitting a scarf, the Big Bang indexed by $t=0$ is the bottom edge of the scarf.

The foregoing picture can be considered a version of J. M. E. McTaggart's so-called "A-series" conception of temporal events, as contrasted with his "B-series." To review this terminology: in a famous paper, McTaggart (1908) presented an argument against the independent existence of time by observing that temporal events need to be characterized in two different (A vs. B) ways. In the "A-series," an event is characterized by whether it is in the past, present, or future; while in the B-series, the same event is characterized only by its position relative to other events (i.e., the usual temporal index t). McTaggart argued that both characteristics are necessary for time to exist, but that (in keeping with the usual idea of a "moving now") any given event E indexed by t will "at different times" be past, present, *and* future. In order to specify "when" to apply those differing A-series descriptions, it then appears necessary to invoke an additional time index, say s. Then, for example, we can truthfully say "event E at time t is in the present when time s is the present." But in order to say when time s is present, we need a third index; and so on, ad infinitum. So, according to McTaggart, statements involving truly temporal properties (e.g., past, present, and future) cannot be unambiguously true or false (because of the infinite regress involved in attempting to pin down their truth or falsity). He concludes, based on the indefinite truth character of statements about time, that there is really no such thing as time.

Now, McTaggart's argument against the independent existence of time is not a problem for PTI, since PTI does not assert that time exists as an independent substance. According to PTI, what is fundamental in the temporal sense is just the now, which changes. So *change* is primary, and the time index is just a way of recording those changes. However, PTI differs from the standard "A-series" account in that there are no

[10] I owe this insight to Wendy Hagelgans.

[11] These "knitting stages" can be thought of as Stapp's "process time" (Stapp, 2011), though he views the "now" as advancing while I differ from that aspect of the picture.

actualized "future" events (future entities are no more than possibilities – in the analogy, they are just yarn and ideas about what to do with the yarn).

Now let us continue with the "knitting" analogy. The domain of classical phenomena can be characterized as a "fabric" in which the stitches are very small, uniform, and tight, and we can think of the laws of motion as predictable colored patterns in the resulting fabric. But, if we "zoom in" on the same fabric (as in the Chapter 7 discussion of "zooming in on a baseball"), and/or work hard to create certain phenomena, the stitches can be seen to be removed from the needles at times, giving rise to patterns of a different character (such as cables that seem to "float" above the background of knitted fabric). These are quantum phenomena, arising in particular in the delayed choice and similar experiments (recall Chapter 5).

In Figure 8.3(a), some stitches are removed from the knitting process and held in an "indeterminate state" (on the cable needle) as (b) surrounding stitches are knitted into the "past" (the extruded fabric). Thus the standard "classical" evolution of the various phenomena continues, except for those "indeterminate" stitches that are held back until a later stage. In (c), the indeterminate stitches are made determinate as they take their place in the fabric. The result is a pattern with more texture and depth than the plain "classical" fabric. Just as in the knitting process, the creation of events is a process of stitching between the past (i.e., the stitches being extruded) and the future (the balls of yarn providing the raw materials). This is a kind of "growing universe" theory of time;[12] but it is the past that grows and continues to become actualized as it falls away from the present (the "now" does not "advance"). Meanwhile, the future is not a realm of determinate events but rather a realm of physical possibilities – the "raw material" for events, if you will. The present does not "advance" into the future. Rather, the future is a set of possibilities that becomes woven into the created past through the action of now.

In previous chapters, and especially in Chapter 6, I discussed PTI in terms of the Wheeler–Feynman and Davies absorber theories of radiation. In those theories, the temporal direction of energy transfer (i.e., positive energy propagating in the positive temporal direction) is based on a future boundary condition. In terms of the above analogy, that future boundary condition is simply the fact that there is a finite amount of yarn, and/or that the chosen pattern dictates the placement of the

[12] As mentioned earlier, the first "growing universe" approach to time was proposed by C. D. Broad (1923). Earman (2008) gives a critical discussion of Broad-type "growing universe" theories. I believe that the transactional model resolves many of the challenges Earman raises for Broad-type theories. For instance: (i) a direction for "becoming" is clearly specified in terms of the distinction between emission and absorption; and (ii) transactions provide the kind of dynamic creation of events that he worries seems to be missing in Broad's original approach. Note that the present model has no problem with future-tensed sentences as outlined in his discussion; the truth value of a future-tensed sentence is indeterminate at the time it is uttered, as argued below in Section 8.5. This is because the model does not consist of a set of ordered "chips off a Newtonian block"; there is genuine indeterminacy in the becoming. (Recall that in this model, the future is real – as a set of potentialities – but not *actual*.)

last row (e.g., the "other end of the scarf"). That is, one does not need an *actualized* event in the future to provide for the necessary future boundary condition; that condition can be provided by a real but possible entity (recall Figure 7.2, category II).

8.1.4 Becoming vs. relativity theory

The above account may raise the worry that a kind of "absolute simultaneity" is being smuggled in, which is at odds with relativity's banishment of that notion. That is, doesn't a "row of stitches" count as a set of simultaneous events? Yes, but only with respect to a given observer, and that does not translate into the claim that events carrying the same time index share the same "now." As Stein notes, "*'a time coordinate' is not time*" in relativity theory (Stein, 1968, p. 16). Recall that anything outside an event's light cone is strongly extra-empirical (meaning that it cannot engage in any transactions with that event). Therefore, as Stein (1968, 1991) argues, it is at best physically vacuous, and arguably inconsistent with relativity theory, to attribute "nowness" to an event outside the light cone.

In view of the above, it should be noted that the "knitting" analogy is only a partial one in that it is not really accurate to assume that becoming literally takes place in a "row," or along a given plane of simultaneity.[13] In this regard, I take note of arguments to the effect that relativity is incompatible with a "becoming" picture of events;[14] but such arguments have depended on (1) a substantivalist notion of spacetime which takes "events" as mathematical points in a pre-existing spatiotemporal "substance," and/or (2) an assumption that the "present" can only be defined with respect to a particular plane of simultaneity, which I have already disputed above. Stein (1968, 1991) makes a persuasive case against assumption (2), even as he uses a substantivalist picture to argue that no event can be (in my terminology) actualized with respect to another event that is spacelike separated from it unless all events are actualized, leading to a "block world."

Under PTI, the fundamental structural component of becoming is the actualized transaction, which establishes only two spacetime "points." (This feature is further elaborated in Section 8.2 below.) In this non-substantivalist picture, one cannot use Stein's argument (Stein, 1991, pp. 148–9) that given two spacelike separated events *a* and *b* such that *b* is actual with respect to *a*, all other spacetime "events"[15] must

[13] This feature is a key distinction between the current proposal and that of Tooley (1997); the latter posits an absolute space and therefore an absolute rest frame, so Tooley's "fabric" does have a horizontal leading edge.
[14] Cf. Putnam (1967), Rietdijk (1966), Penrose (1989). These are referred to as "chronogeometrical fatalism" by Savitt (2008).
[15] "Events" in quotation marks refers to "unoccupied" spacetime points in a substantivalist approach.

also be actual with respect to *a*, including those in the future with respect to *a*. That argument requires that spacetime be a mathematical manifold of not necessarily occupied points "events," which I deny.[16] Recall also that the actualization of a transaction is an a-spatiotemporal process; it is the *coming into being* of an event (actually two linked events, the emission and the absorption; but in general the emission event can be identified with an already established transaction or set of transactions, such as the existence of a macroscopic source of offer waves). *The actualization of a given transaction is the "now" for the associated absorption event.* In that sense, I agree with Stein's view that "now" is properly defined only with respect to particular events; it is not appropriate to consider "now" as applying to an entire manifold of events "at the same time," since the latter phrase smuggles in an inappropriate simultaneity notion, as well as contradicting the a-spatiotemporal nature of the actualization process.

The motivation underlying much of the "block world" vs. "becoming" debate is the desire to see the "becoming" from a "God's-eye view"; that is, from "outside" the spacetime structure that is "growing." This approach implicitly (or explicitly) invokes an additional pseudo-temporal parameter characterizing the "growing" (and this, too, is where our "knitting" analogy falls short). Stein eschews this approach, and holds that the only sense in which the "present" can be defined is with respect to individual events. With regard to an observer's subjective perception of "now," the "present" or the "now" is always a local phenomenon. Significantly, the French word for "now" is "maintenant": literally "holding in the hand." The elusive nature of "now" comes from the fact that it is necessarily a non-collective property; it applies to each individual transaction's a-spatiotemporal actualization process. In Stein's terms, an event is only present to itself; "in [relativity theory], the present tense can *never* be applied correctly to foreign objects" (Stein, 1968, p. 15). He goes on to express a view of becoming that applies to the present model (with the modification that instead of the "spacetime point" referred to below, what exists is an actualized absorption event):

In the context of special relativity, therefore, we cannot think of temporal evolution as the development of the world in time, but have to consider instead (as above) the more

[16] This is a version of "chronogeometrical fatalism." My version of Stein's "*Rab*" relation, which says "*b* has become with respect to *a*," therefore need not limit *b* to the past light cone of *a*. My model would also appear to be immune to a similar argument of Weingard (1972), since his is also based on a substantivalist view of spacetime and the assumption, based on the conventionality of simultaneity (i.e., choice of the one-way speed of light), that *any* "event" outside the light cone of an actualized event must correspond to an actualized event. I go in the opposite direction, in a sense: one cannot assume that any "event" outside one's light cone corresponds to an actualized event. Such an "event" corresponds to an actual event *only* if it is the absorption or emission site of an actualized transaction, not by reference to the structural features of a pre-existing spacetime substance (which I deny). Along with this would go the requirement that an actualized event not be in the future of any other actualized event. But the set of actualized events is contingent on the actualization of specific transactions, not the structure of a spacetime substance, so there is no inconsistency.

complicated structure constituted by . . . the "chronological perspective" of each space-time point. (p. 16)

Another account of the locality of the empirical "now" is given in Norton (2010):

> The "now" we experience is purely local in space. It is limited to that tiny part of the world that is immediately sensed by us. There is a common presumption of a present moment that extends from here to the moon and on to the stars. That there is such a thing is a natural supposition, but it is speculation. The more we learn of the physics of space and time, the less credible it becomes. For present purposes, the essential point is that the local passage of time is quite distinct from the notion of a spatially extended now. The former figures prominently in our experience; the latter figures prominently in groundless speculation. (p. 24)

In the present model, rather than a "passage of time," we have the generation of an ever-increasing "fabric" of past events, but the basic observation is the same: "now" is a local phenomenon.

8.1.5 The "dead past"

Here I address another issue that is at play in contemporary discussions of "growing universe" pictures of time: namely, how to understand the "dead past" feature of the model I propose here. That is, "people in the past" (such as Socrates) are not observers having empirical experiences. The actualized past is like the cast-off skin of a snake; the living, experiencing snake is no longer contained in it.

This model is in contrast to "presentism," the view that only the present exists. Heathwood (2005) argues, in response to Forrest (2004) whose model is similar to this one, that regarding people in the past as non-conscious leads to the same problems plaguing "presentist" accounts of time. The problem for presentism is that there seems to be no plausible way to account for the meaningfulness of a statement such as "I admire Socrates" if Socrates, being in the past, does not exist. To what, then, does the sentence refer? The growing universe approach sidesteps this problem, since in that approach, Socrates *does* exist in the past. However, Heathwood argues that the same problem reappears in the "dead past" version of the growing universe for sentences such as "Socrates was conscious when he was killed." He certainly has a valid point if such sentences are taken as referring to the "dead past." But I would argue that the referents of such sentences are just earlier stages in the process of the growing universe. So the referent for the above sentence is the stage at which Socrates' execution was "on the needle," or "in the present (or, taking into account the previous section, when Socrates was "present to himself").[17] I see no reason why such statements cannot refer to an earlier stage in the process; language need not be limited to any particular stage.

8.2 Spacetime relationalism

I noted in Chapter 4 and above in Section 8.1 that PTI assumes a relationalist (or "antisubstantival") view of spacetime. In this section, I examine relationalism in more detail, following aspects of the formulation in Friedman (1986).

8.2.1 Relationalism vs. "substantivalism" about spacetime

The spacetime realist (or "substantivalist") views spacetime as a substantive manifold M of points $\{a, b, c, \ldots\}$, each indexed by the temporal coordinate t and spatial coordinates (x, y, z), where all the indices are real numbers ranging from minus infinity to plus infinity. The manifold itself is considered to have structure in the form of symmetries and a metric. In particular, according to relativity theory, the square of the spacetime interval ds is a real-valued function $I(a, b)$ defined on M. The key point is that according to substantivalism, not all of the spacetime points correspond to physical events; rather, only those points belonging to some subset P of M are occupied by concrete physical events.

In contrast, the relationalist thinks that there is no substantial spacetime manifold M but that there are *only* concrete events whose collective features contain all the necessary qualities to account for the observed symmetries and phenomena normally associated with spacetime itself. While this work does not attempt to present a case for relationalism (that has been ably provided by numerous authors[18]), it seeks to place PTI in the context of the discussion concerning the competition between these two views. As Friedman (1986) has noted, relationalism has no significant challenges in accounting for the symmetry aspects of spacetime; indeed, it has advantages over the substantivalist view, but the consensus among researchers is that it runs into a significant challenge in accounting for the laws of motion. Put simply, this is because those laws seem to require an absolute background of some kind for their formulation, and this is where M appears indispensable. Later on in this chapter we will see that the basis of actualized events in transactions provides for the possibility of an unambiguous basis for the laws of motion.

[17] The same referent would apply to the sentence viewed as unproblematic by Heathwood, "Socrates was fat when he was killed." So this also resolves the concern he raises about inconsistency of the "truth-makers" for the two types of statements.

[18] For example, Barbour (1982), Brown (2005), Earman (1986), Sklar (1974).

8.2.2 Distinct facets of relationalism

Friedman distinguishes two main facets of relationalism in the literature: what he terms *ontological* and *ideological*. They are defined as follows:

"Ontological": Spacetime is no more than the set of existing events *P*.
"Ideological": Existing spacetime events meet certain physical requirements (such as "causality").

These two approaches are not different versions of relationalism but rather aspects of it that are primarily under debate. For example, as Friedman notes, "ontological" considerations were primarily at issue in the Newton–Leibniz debate, which concerned Newton's postulation of an "absolute space" and "absolute time" held by Leibniz to be without legitimate physical content; while "ideological" considerations have been at issue in the more recent discussion revolving around Reichenbach's and Grünbaum's contributions.[19] PTI's relationalism could be said to address primarily the "ideological" aspect in that it defines eligibility for membership in *P* (the set of concrete events) in terms of a specific physical process: the transaction.

As noted previously in this work, most practicing physicists believe very strongly in spacetime as a substance; that is, as an entity that exists in its own right as a dynamic "container" which supports events and influences their interactions. Yet spacetime itself is not observable. There is no actual empirical evidence of the independent existence of a spacetime substance or "substratum," as something distinct from events. Rather, the existence of spacetime is *inferred* from observable phenomena based (in large part) on the metaphysical view that events require a "container"; i.e., the view that it is not enough to say just that events themselves (and their collective structure) exist. Perhaps the strongest theoretical support for the notion of a spacetime substance is found in general relativity, which relates the metric characterizing sets of events to the mass–energy of the fields comprising the events. But this can be understood in terms of a relational (antisubstantivalist) view of spacetime (cf. Brown, 2002, p. 156). The basic point is that spacetime is an extra-empirical notion, in the same sense that causality is an extra-empirical notion: neither is actually observed, nor directly referred to by theoretical entities.

Now, the reader might protest: "Surely, spacetime *is* referred to because many entities, such as fields, contain spacetime arguments: for example, $\Psi(x, t)$." However, there are (at least) two reasons why the use of such arguments does not constitute a reference to spacetime substance: (i) the arguments (x, t) are not invariant, i.e., they are dependent on the state of motion of the observer; and (ii)

[19] Cf. Reichenbach (1958), Grunbaum (1973).

they are defined only relative to distances (intervals) between events, or to an arbitrary coordinate system, not in an absolute sense. Both (i) and (ii) imply that (x, t) refers to *a relationship between events* (and/or observers) rather than to something external to those events. Moreover, strictly speaking, the arguments in $\Psi(x, t)$ index *points on the field* Ψ rather than spacetime points.[20] (While practitioners of quantum field theory often characterize the theory in terms of the association of a field with "all points in space," that formulation simply adds to the theory the presumption that space is a pre-existing substance, while the latter is not a necessary component of the theory itself.)

The remainder of this chapter describes how transactions create an interlocking, structured set of events which can do the work of "spacetime" without the necessity of invoking a background spacetime "substance."

8.3 The origin of the phenomenon of time: de Broglie waves

While the relational, antisubstantivalist view espoused herein rejects the idea that time is a substance or pre-existing entity, we can nevertheless point to a "primal clock" whose origin is in the fields (offer and confirmation waves) giving rise to transactions (actual events). This primal clock is simply the de Broglie wave pulse corresponding to the frequency of the quantum associated with it (recall Section 6.3.6). First, some preliminary remarks are in order concerning the relationship of de Broglie waves to the spacetime construct.[21]

8.3.1 The de Broglie phase clock

The best way to investigate this issue is to revisit de Broglie's 1923 dissertation. Recall that de Broglie's hypothesis was that matter, as well as light, had a periodic aspect associated with it. He did not pretend to know what this periodic aspect was in a material sense, but simply analyzed its formal properties. In his words:

> ... it seems to us that the fundamental idea pertaining to quantum is the impossibility of considering an isolated quantity of energy without associating a particular frequency to it. This association is expressed by what I call the "quantum relationship," namely:
>
> $$\text{energy} = h \times \text{frequency}$$
>
> where h is Planck's constant ... The notion of a quantum makes little sense, seemingly, if energy is to be continuously distributed through space; but we shall see that this is not so.

[20] This point is made in Auyang (1995, p. 48).
[21] This section is loosely based on Kastner (2011a). The final publication is available at www.springerlink.com.

One may imagine that, due to a meta-law of Nature, to each portion of energy with proper mass m_0, one may associate a periodic phenomenon of frequency ν_0, such that one finds

$$h\nu_0 = m_0c^2$$

The frequency ν_0 is measured, of course, in the rest frame of the energy packet . . . Must we suppose that this periodic phenomenon occurs in the interior of energy packets? This is not at all necessary; [section 1.3 of his dissertation] will show *that it is spread out over all space* . . . An electron is an atom of energy not in virtue of occupying a small volume of space, I repeat: it occupies all space, but rather in virtue of its indivisibility, in which it constitutes a unit. (de Broglie, 1923, pp. 7–8; English translation,[22] emphasis added)

PTI is in agreement with de Broglie's view that energy associated with a quantum is not localized in a "packet." But from the perspective of PTI, which denies the fundamentality of "spacetime," his remark that the energy is "spread out over all space" should not be taken literally but instead interpreted as follows. "Spacetime" is just a map that we draw to coordinate our perceptions: it is strictly a phenomenal notion. Entities such as offer waves that are not part of actualized transactions cannot be associated with any particular spacetime region; they *transcend* the spacetime construct, and it is in that sense that they are "non-local." But under PTI they are physically real, so if we want to represent them somehow for pragmatic purposes on a spacetime map, we have to depict them as "spread out over all space."[23]

In response to the possible objection that an entity being "spread out over all space" means that it has to be "in spacetime," I note with Descartes (recall Section 2.4) that if something is *everywhere* (or rather, in PTI terms, representable on the spacetime map only as occupying the entire map) then it is entirely unobservable. Such a situation is empirically indistinguishable from that same entity being *nowhere*. For example, one can think of this as a color overlay on a map of the world, which affects everything on the map the same way for all locations and times and for all observers and is therefore unobservable because it can effect no change in any measuring or perceiving entity. Alternatively, it is just the same as drawing the map with a different color ink and/or using a different color paper: *it is the same map*. Thus, entities that are not "actualized" according to PTI and which thus transcend the level of appearance (in Platonic terms) can be considered either "everywhere" or "nowhere" with respect to spacetime. In either case, we are projecting an entity whose existence is not confined to any spacetime region onto

[22] This excerpt's English translation has been slightly modified in places by the author for clarity.

[23] This is the same characterization that our game avatars Jon and Mia (of the previous chapter) would apply to the game "table-in-itself" created by Jonathan: from their perspective, the abstract information giving rise to the appearance of the game table is non-local but physically efficacious, so from their perspective it might just as well be "spread out over all space."

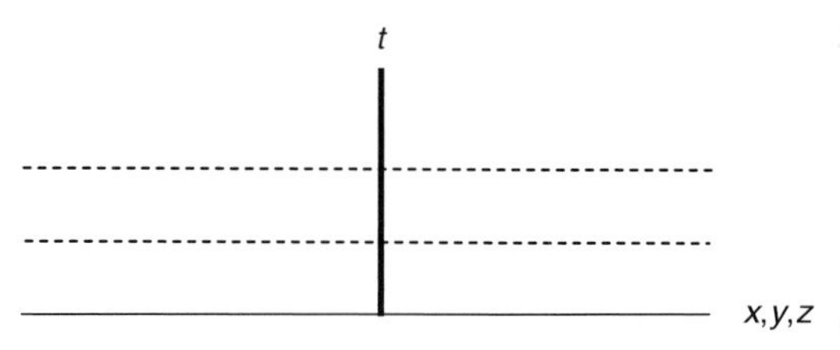

Figure 8.4 The pulses of the de Broglie phase wave "clock" can serve to establish non-local hyperplanes of simultaneity (horizontal lines represent 3-dimensional spatial hyperplanes).

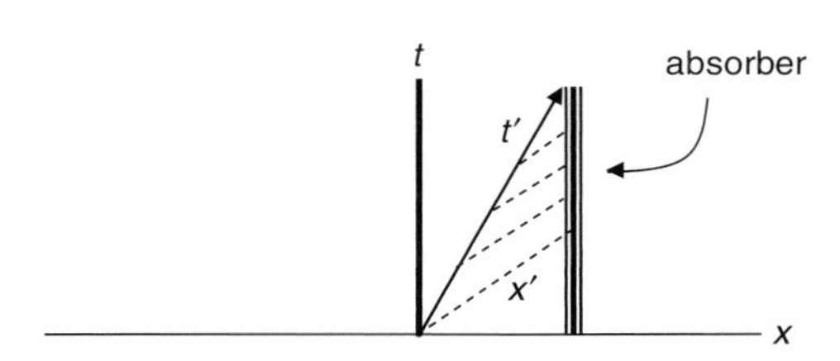

Figure 8.5 A transaction establishes a group velocity as seen from a frame of reference moving with respect to the quantum of Figure 8.4. The quantum's proper time is t' and its axis of simultaneity is x' whose (reciprocal) slope corresponds to its phase velocity u. (Untransacted offer wave component heading to the left is not shown.)

a spacetime map which is strictly an empirical-level construction and which can represent without distortion *only* the level of appearance.

Thus PTI takes seriously de Broglie's proposal that quanta are explicitly non-local and associated with a sub-empirical periodic phenomenon.[24,25] The latter is just the phase wave, discussed earlier in Chapter 6, which establishes the spatial axes (lines of simultaneity) for the quantum. This is illustrated in Figure 8.4. It should be noted that the origin is completely arbitrary.

8.3.2 The group wave and the temporal axis

The offer wave (de Broglie phase wave) acquires a group velocity only upon actualization of a transaction, as shown in Figure 8.5. (We can represent a macroscopic absorber on the spacetime map because it has been localized via actualized transactions among its atomic constituents, our laboratory environment, and our

[24] That is, the pulse does not exist "in spacetime" with respect to any pre-existing temporal axis, just as the spin of an electron does not exist in spacetime and is generally regarded as taking place in an "abstract" or "internal" space.

[25] Dolce (2011 and references therein) has also explored de Broglie waves as the basis for spacetime, though his approach is not time-symmetric.

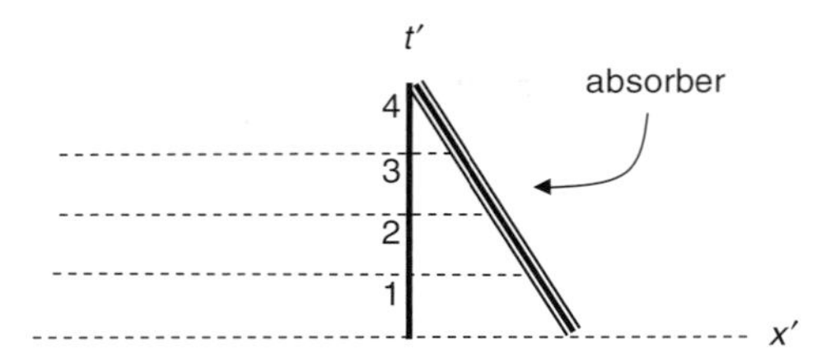

Figure 8.6 The same process as viewed from the quantum's rest frame. As a result of the transaction between the emitter and the absorber, an observable mass m_0 comes into being at $x' = 0$.

sense organs.) The (reciprocal of the) slope of the quantum's time axis reflects the group velocity v_g, which is the usual velocity we would associate with a particle moving from the emitter to the absorber. Figure 8.6 shows the same process as viewed from the rest frame of the quantum. In this perspective, the absorber moves toward the quantum, and a transaction occurs in which the quantum is actualized but has not traveled any distance; $v_g = 0$.

Thus, in the PTI picture, the "meta-law" referred to by de Broglie is the principle that matter/energy can only be manifested (transferred) in a confirmed offer wave (de Broglie wave); and it is only a confirmed, actualized wave which reflects the group velocity corresponding to the spatial momentum of the quantum transferring the energy. The manner in which the phase wave of frequency ν is "associated" with energy is that it provides the *physical possibility* for a transfer of energy in the amount $h\nu$. The non-local nature of the phase wave is a reflection of the fact that it is not contained in spacetime: it is a pre-spacetime entity.

The rest or "proper" mass m_0 of the particle determines its primal clock frequency. *This suggests that the origin of time is contained in matter* (just as no stitches can be extruded if there is no yarn placed on the knitting needle). In a relational account of spacetime, without matter there is no primal clock and therefore no time. Recall that a "free" photon (an idealization as discussed in Chapter 6), having zero rest mass, does not "see" any passage of time (or, for that matter, any spatial separation). Its propagation is restricted to what is termed a "null cone"; corresponding to spacetime intervals of zero, where the interval ds is defined by $ds^2 = dt^2 - dx^2 - dy^2 - dz^2$. According to PTI, if there were no matter, there would be no spacetime. It is matter that *creates* the separation between temporal and spatial axes through transactions establishing group waves (time axes) with respect to the hyperplanes of simultaneity defined by the phase wave clocks (see Figure 8.7).

Regarding photons, however, it is important to recall the discussion in Chapter 6 concerning the somewhat misleading distinction between "real" and "virtual" photons. A "real" photon is an idealization. If a photon is emitted and absorbed, it

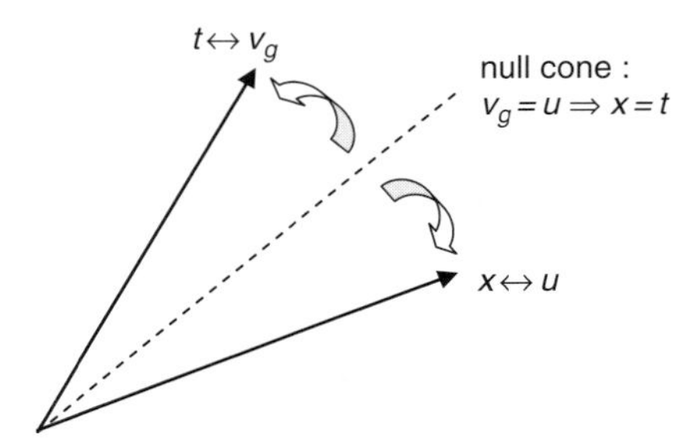

Figure 8.7 An actualized transaction involving a quantum of matter (i.e., a quantum of a field with non-zero proper mass) is the basis for the separation of spatial and temporal axes through the difference between the phase and group wave velocities of the quantum. (Only one spatial dimension is shown.)

is technically a "virtual" photon, which means that it can have non-zero proper mass. Thus any *actually detected* photon can be said to traverse a finite (even if *almost* null) spacetime interval, i.e.: $dt^2 \approx dx^2$ (in one dimension). The key point is that spacetime intervals are no more, and no less, than measures of realized transactions resulting in transfers of energy from an emitter to an absorber, which themselves are pre-spacetime objects at the micro-level. Energy is only transmitted by way of the group wave which exists only in the case of an actualized transaction.

This picture is consistent with – and arguably a better explanation for – the theoretical fact that processes involving "virtual" quanta violate energy conservation. Conventionally, this fact is understood in terms of the uncertainty principle: it is argued that energies in violation of the conservation laws involve virtual quanta with short enough lifetimes Δt that a violation in the amount ΔE of energy satisfies the uncertainty principle: $\Delta E \Delta t \geq h$. However, it is not really legitimate to assume that virtual particles have a "short" (or even definable) lifetime; as noted by Feynman, any particle for which one takes into account its emission and absorption is technically a virtual particle.

In the PTI picture, virtual particle energy conservation violation can be understood very simply: *physical quantities (such as energy and momentum) are only conserved at the level of actualized transactions* because they are only *possible, sub-empirical* quantities in the absence of actualized transactions. Virtual particle processes (such as those discussed in Chapter 6 as internal lines in scattering interactions) occur *only at the level of possibility* and can therefore violate energy conservation. This is also an answer to the question: "where did the energy (or momentum or angular momentum) go?" in all the unrealized components of the wave function that do not correspond to actualized events. The answer is that it was never empirically present to begin with – it was only *possible* energy, etc.

Yet another aspect of the inconsistency of quantum-level processes with long-standing empirical-level physical principles is found in the fact, as noted in Brown (2005), that quantum test particles do not obey the "zeroth law of mechanics," i.e., the principle that "the behavior of free bodies does not depend on their mass and internal composition" (Brown, 2005, p. 25). This is easily seen by looking at the time-dependent Schrödinger equation, which depends explicitly on the mass of the quantum. Again, this discrepancy can be understood by considering quantum mechanics as describing the behavior of sub-empirical objects that do not have to obey empirical-level principles of mechanics.

8.3.3 Why acceleration is absolute

As noted in Section 8.2, it is often argued that a relational account of spacetime cannot explain why acceleration has a different status than inertial motion: the former is absolute, while the latter is relative. Acceleration is absolute in the sense that we cannot "transform it away"; i.e., obtain a non-accelerated but otherwise equivalent physical account by describing the set of events with reference to an observer in a different state of motion. The accounts will be inequivalent in that forces will be present in one that are not present in the other. Thus, the substantivalist's argument goes, there really must be a "spacetime" with respect to which a system is accelerating. Another way of presenting the argument is that one cannot even formulate laws of accelerated motion without appeal to special spacetime entities such as the inertial frames defining a state of non-acceleration. These may not be occupied by concrete events, and therefore are not available to the relationalist for which "spacetime" is no more than the set of existing events.

It is the transactional process that endows accelerated motion with its special status, eliminating the need for reference to an independently existing spacetime substance. Recall that in PTI, the basic empirical physical phenomenon is an actualized transaction. As noted above, a transaction is what defines (in stronger ontological terms, *creates*) a spacetime interval and a transfer of momentum of a constant value as described above in terms of the de Broglie group wave. The value of this momentum depends on the state of motion of the observer, but it is constant and rectilinear because it involves a single transaction with only two spacetime points; it cannot reflect any acceleration. In order to establish acceleration, one needs at least three spacetime points (two intervals), as I discuss further below. (To be clear, in referring to "two points," I do not mean two pre-existing spacetime points. The interval corresponding to the emitter/absorber transaction is primary, and it is *in virtue of* that actualized interval that we can identify two endpoints which now qualify as different spacetime events.)

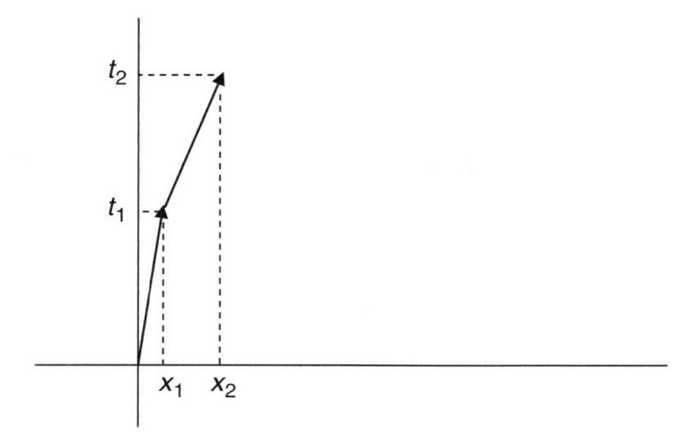

Figure 8.8 Acceleration requires at least two transactions and therefore differs from inertial motion (needing only one transaction) in an absolute sense, but without reference to a substantive spacetime.

Acceleration is defined as the change in velocity per unit time:

$$a = \frac{dv}{dt} = \frac{d}{dt}\left(\frac{dx}{dt}\right)$$

In PTI, time and space are indexed only by endpoints of actualized transactions, so these indices are therefore discrete (and see Figure 8.8):

$$a = \frac{\Delta v}{\Delta t} = \frac{\Delta}{\Delta t}\left(\frac{\Delta x}{\Delta t}\right) = \frac{(x_2 - x_1) - (x_1 - x_0)}{(t_2 - t_1)(t_1 - t_0)}$$

where we can establish a "unit time" only by reference to some other transactional situation which provides regular pulses (such as an atomic clock). There are two spacetime intervals ds involved here: $ds^2(1,0) = (t_1 - t_0)^2 - (x_1 - x_0)^2$ and $ds^2(2,1) = (t_2 - t_1)^2 - (x_2 - x_1)^2$. So, while in order to define velocity we need only two spacetime points, we need at least three to define acceleration. Thus acceleration differs from inertial motion in an absolute sense, by requiring more than one transaction.

Acceleration due to a force field can be understood as a sequence of transactions between an offer wave and the force-mediating quanta of the field. Thus, in synchrotron radiation, an electron radiates due to its interaction with an external magnetic field that serves as a "pump" inputting (effectively) free photons, which induce emission processes (transactions) between the accelerating electrons and detectors. In contrast, as discussed in Chapter 7, an electron bound within an atom does not radiate because no external transaction is available to it. Such an electron is not really accelerating; this is why its state is called a "stationary state." The electron + nucleus of the atom constitute a bound state which itself is a more complex offer wave: i.e., a unified system that is held together by its associated potential energy. One has to put energy into that system before the electron can access possible external transactions (i.e., for the electron to be liberated from the atom). To sum up the argument in this subsection: acceleration only occurs for phenomena involving more than one transaction in which momentum transfer changes its value, and therefore differs from non-accelerated motion in an absolute sense, without reference to any background substantive spacetime.

8.4 PTI vs. radical relationalism

PTI, while eliminativist about time as a substance, provides a useful compromise between the radical relational view (which I will term "RR") of Carlo Rovelli (1996) and the substantival view of time (i.e., that time is a real entity). Rovelli's picture follows the quantum state through its progressive interactions with measuring devices in the usual way, in which the linearity of the state carries over to macroscopic processes because absorption – which could break the linearity via confirmations and the triggering of transactions – is neglected. Thus, in Callender's words:

> Consider the famous case of Schrödinger's cat. The cat is suspended between life and death, its fate hinging on the state of a quantum particle. In the usual way of thinking, the cat becomes one or the other after a measurement or some equivalent process takes place. Rovelli, though, would argue that the status of the cat is never resolved. The poor thing may be dead with respect to itself, alive relative to a human in the room, dead relative to a second human outside the room, and so on . . . It is one thing to make the timing of the cat's death dependent on the observer, as special relativity does. It is rather more surprising to make whether it even happens relative, as Rovelli suggests, following the spirit of relativity as far as it will go. (2010, p. 64)

This is indeed the logical conclusion of applying a relational view to quantum mechanics if one does not take absorption into account. PTI differs both from RR and from the "usual way of thinking" regarding non-relativistic quantum mechanics, in that it provides for a definite empirical result to occur: i.e., an event is actualized (via a transaction) in which the cat is definitely alive or definitely dead. Under TI, the measurement process is clearly defined as explained in Chapters 3 and 4: it is precipitated by absorption. Since there are so many absorbers in a macroscopic situation, the measurement is completed long before any macroscopic object could be placed into a linear superposition. Note that this situation is still perfectly consistent with relativity in the sense that the spacetime coordinates given to the actualized event are relative to an observer. Two observers in different inertial frames will disagree on the coordinates of the event, but will agree on the spacetime interval between that event and another event (and on what those events are). In PTI, as opposed to RR, specific empirical events do exist; it is only their individual spacetime "location" which is relative, which reflects the fact that spacetime does not exist as an independent entity.[26] In this regard it is useful to consider Callender's apt analogy between time and money[27] as secondary, derivative notions:

> In Einstein's thought experiments, observers establish the timing of events by comparing clocks using light signals. We might describe the variation in the location of a satellite around earth in terms of the ticks of the clock in my kitchen, or vice versa. What we are doing is describing the correlations between two physical objects, minus any global time as intermediary. Instead of describing my hair color as changing with time, we can correlate it with the satellite's orbit. Instead of saying a baseball accelerates at 10 m/s, we can describe it in terms of the change of a glacier. And so on. Time becomes redundant. Change can be described without it . . . This vast network of correlations is neatly organized, so that we can define something called "time" and relate everything to it, relieving ourselves of the burden of keeping track of all those direct relations . . . But this convenient fact should not trick us into thinking that time is a fundamental part of the world's furniture. Money, too, makes life much easier than negotiating a barter transaction every time you want to buy coffee. But it is an invented placeholder for the things we value, not something we value in and of itself. Similarly, time allows us to relate physical systems to one another without trying to figure out exactly how a glacier relates to a baseball. But it, too, is a convenient fiction that no more exists fundamentally in the natural world than money does. (p. 65)

Interestingly, under PTI, what quantum mechanics in fact does is to "negotiate a barter transaction" every time a quantum is emitted and absorbed. While it is too much trouble for us to keep track of (as Callender notes), nature performs this complicated bookkeeping task admirably, which is why the vast network of correlations is so "neatly organized." The events themselves are actual for everyone; it is only their spacetime descriptions that are relative. Therefore, while I admire the spirit of Rovelli's exploration, I think it is not necessary to deny that clearly defined events exist. Relationalism need not deny that specific events exist; it need only deny that spacetime exists. Indeed the core of relationalism is that it is the structure of the collection of events that defines what we think of as "spacetime."

8.5 Ontological vs. epistemological approaches, and implications for free will

The metaphysical picture proposed here may seem strange or "far-fetched."[28] But there is nothing inconsistent about it and much to recommend it as a viable ontology

[26] While I focus here on the unreality of time, the basic relational view is that the spatial component of spacetime is non-fundamental as well.

[27] Hence the equivalence often cited between the two: they are both equally illusory.

[28] Of course, "many worlds" interpretations can certainly be considered at least as "far-fetched," so one should be careful to avoid a double standard here. We should also keep in mind that it was considered far-fetched for Galileo to insist that the earth was in motion when any one of his contemporaries could clearly "observe" that it was not moving. Appearances can be deceiving.

underlying quantum theory. If one takes the mathematical objects such as state vectors as referring to something ontologically real (as opposed to being just a measure of our knowledge or ignorance), then the entities and processes to which they refer obviously cannot "fit into" spacetime and therefore, if real, must exist in some pre-spacetime realm (PST). This, again, was noticed even by Bohr in his previously quoted comment (see Section 2.5) that such processes must "[transcend] the frame of space and time."

Nevertheless, one might consider whether we should resort instead to an epistemic-type interpretation of quantum states, as in the time-symmetric "hidden variables" approach of Price (cf. 1996) or the models studied by Spekkens (cf. 2007).[29] Such interpretations imply a "block world": i.e., that events are already "there" in spacetime and that various types of "hidden variables" (i.e., unknown aspects of the "ontic state" of a system) encode additional information about which events, out of an *apparent* choice among possible events, are actually "chosen already."

I believe that such an approach – taking quantum theory as "incomplete" – misses a valuable opportunity to discover what truly novel message might be contained in quantum theory, as discussed in Chapter 2. For one thing, such approaches (e.g., the Bohm model) have ongoing difficulties with the relativistic domain, while PTI is fully compatible with it.[30] Moreover, there would appear to be no room in the "block world" implied by a hidden variables approach for the experience of human agents as having free will concerning what they choose to measure or to create; there is no *genuine* becoming. Of course, this brings us back to the age-old philosophical discussion concerning fatalism versus free will, and I do not pretend to do justice here to this intricate and never-ending debate. However, it is generally accepted that in a strictly predetermined world, there can be no genuine free choice in the sense of an unconstrained selection of one path from a "garden of forking paths," since there are no forking paths. In order to "save the experience" of free will, one has to resort to an argument that one can "freely" choose what one is already destined to choose.[31] If in fact choices are already made and already there in the block world, then there are no real "choices" at all, and our perception that we are really making free choices is an illusion.

[29] As noted in Chapters 1 and 2, Spekkens' models may, in any case, run foul of the Pusey *et al.* theorem (2011).

[30] Sutherland (2008) proposes a time-symmetric version of Bohm's theory, which does have basic compatibility with relativity. This constitutes another type of epistemic, block world-type interpretation.

[31] This position is known as "compatibilism," the view that determinism is compatible with free will. While much of the free will vs. determinism debate concerns moral responsibility and is therefore beyond the scope of this work, the basic compatibilist argument boils down to the idea that free will just means being able to act in accordance with one's wishes in an unfettered manner. Given determinism, one's wishes must, of course, be fully determined. This does not address the situation in which an experimenter has an apparent choice between two possible measurements but no personal preference of one over the other; one cannot mitigate the force of determinism against free will by appealing to one's wishes or desires in such cases.

Nevertheless, a recent trend seems to be emerging among some interpreters of quantum theory: the idea that physics implies that there is a block world, and that the correct interpretive task should therefore be to examine the ramifications of that ontology. As noted above, I believe that this is a mistake based upon taking a particular kind of map for the actual territory. If one believes that the block world model is correct, one consequence that follows is that the events that we see unfolding around us as we "move along our worldlines" don't actually "happen" in any particular order; i.e., that all events simply exist in the block world and that therefore the direction of events is arbitrary and a matter of perspective. A block world adherent might assert, for example, that the directional quality of events is simply a matter of the kind of creatures we are – i.e., that some other kind of creature in the block world would see things entirely differently. A fictional example might be Merlin the Magician, a different kind of creature who is facing in the opposite direction and moving along his worldline in the opposite direction.[32] One can even imagine picking up the "block world" and turning it sideways by ninety degrees; i.e., interchanging space and time (which disregards the important metrical distinction between the space and time indices, and the fact that there is no quantum mechanical time operator while there is a position (space) operator).

Along with the block world approach goes the assumption that many of our "intuitions" about the world must be inaccurate. Among these are: (1) our experience of only one event at a time; (2) the perception of "nowness"; (3) the perception that (classical) radiation proceeds *from* an emitter *to* an absorber in a diverging spherical wave; and (4) the sense that we have free will; i.e., the capacity to intervene in events and alter their future courses by our choices. However, it should be noticed that at least some of these "intuitions," e.g., (1) and (3), are in fact much more than mere intuitions: they are well-corroborated empirical observations.

As is evident at this point, I disagree that one need take the block world as the message of physics. Granted, we may need to revise some of our "intuitions," such as the idea (discussed in Chapter 2) that we can see everything that exists. However, we need to be careful that in jettisoning what might be called "intuitions," *we are not actually jettisoning the empirical reality that physics is supposed to be explaining.* For surely the world of appearance, as reflected in much of the list of perceptions above (but perhaps not including free will), *is* the empirical realm. Since there is an interpretation of physical theory (the one I'm proposing herein) which can explain, rather than jettison, many aspects of empirical reality, surely that is methodologically preferable to taking one kind of map as the actual territory and embracing the

[32] It is important to note that Merlin would be in category III of Chapter 7's possibility types; i.e., he is no more real than "that possible fat man in the doorway."

consequence that we are radically and collectively mistaken about our thoroughly corroborated empirical observations, such as the direction of radiation flow.

The point is that a less radical option is available: simply admit that there may be sub-empirical entities that we are not able to perceive at the empirical level, and that those are the objects to which quantum theory refers. In fact there is ample precedent – i.e., atomic theory, so despised by Mach the staunch empiricist, but subsequently vindicated by its fruitfulness – that this is the best option.

In contrast to interpretations that take quantum theory as referring to observers' ignorance concerning already established spacetime events, PTI does not have to sacrifice genuine free will or make do with an impoverished "illusory free will" substitute, because it is fully harmonious with free will. In PTI, no spacetime events exist apart from actualized transactions. So, for example, the fatalist argument rehearsed in Dummett (1964) does not apply. Dummett's challenge concerns statements about the future. He argues that such statements must refer to something in order to be either true or false, and that their referents are future events, which must therefore exist. However, in the current proposal, sentences about the future such as Dummett's example, "I will be killed in the next air raid," do not refer to spacetime events; they refer only to possible events. Such statements are genuinely neither true nor false because they refer not to pre-existing events but only to possibilities in the PST – events which are *objectively* uncertain.

Note that the above response of PTI to fatalism is not reducible to the claim that the sentence above is neither true nor false at the time it is uttered but will become either true or false at the time to which it refers, in response to which the fatalist can just rephrase the above sentence as "The *statement* about my being killed in the next raid will either become true or false." This is because *both* statements refer only to possible events in the PST, not to actualized events in spacetime. That is, the sentence "The statement about my being killed in the next air raid will either become true or false" is just as much a statement about an objectively indeterminate future as is the original statement, "I will be killed in the next air raid." Both statements ultimately refer to *both* sets of alternative possible events in PST: a subset in which I am killed and a subset in which I am not killed (assuming there actually will be an air raid . . . which is also uncertain!). Thus PTI can deny fatalism while retaining the meaningfulness of statements about the future in terms of real, but objectively uncertain (i.e., unactualized), events.[33] There is no "fact of the matter" about

[33] Dummett (1964) expresses skepticism that one can deny bivalence (i.e., either truth or falsity; no "middle" option) for future tense statements (in fact, he characterizes the response necessary to avoid the fatalist argument as a denial that there can be a "genuine" future tense). But that particular exposition presupposes classical notions about spacetime which one should be prepared to re-evaluate in the face of quantum theory. Moreover, one can question the implicit premise of passivity contained in the air raid example and other examples used to argue for fatalism. If there is genuine free will, then creatures with free will (such as humans) can actively participate in the "weaving" process that is the creation of spacetime events. So, even if one wants to keep bivalence (truth or

whether I will be killed tomorrow when I am making statements (or statements about statements) about tomorrow's air raid.[34] In PST, all statements about the future are meaningful but objectively uncertain because that to which they refer is objectively uncertain.

In the view of this author, taking quantum theory as "incomplete" leaves us with a rather impoverished ontology in which humans must be radically and collectively deceived about their ability to choose and to create. The advantage of PTI over a static block world view is that the networks of transactions *retain* a kind of crystalline beauty sometimes attributed to the block world: transactions certainly express relevant spacetime symmetries; so one can still have the esthetically appealing symmetries without sacrificing a thoroughgoing realist approach that provides additional richness to the ontology, rather than (as in the block world picture) *subtracting* ontological content by denying that the state vector fundamentally refers.[35]

PTI accounts for the empirical spacetime realm in terms of actualized transactions while providing a straightforward basis for subjective experience and free will in terms of a pre-spacetime realm of dynamic possibilities. The connection with the mental realm is not obligatory; PTI is agnostic concerning a relationship, if any, between those possibilities and mental activity. But if there is an empirically unobservable realm transcending the spacetime realm of appearance, that would seem to be a prime candidate for future research concerning a possible connection between subjective experience and quantum theory.

falsity) about future events, one can meaningfully talk about such future events as objectively uncertain but as definitely taking place or not "when the time comes": e.g., I freely may or may not choose to bring about a particular event; but if I do, it definitely occurs, and if I do not, it definitely does not occur. The fact that it ultimately either occurs or not does not mean that my fate was "sealed" at any time prior to that event's actual occurrence (or non-occurrence).

[34] A variant on the block world view is an indeterministic block world, but this is subject to fatalism based on the basic block world assertion that there must be a fact of the matter about any statement about the future.

[35] In case one might argue that "adding richness to the ontology" runs foul of Occam's razor (OR), my response would be that OR applies to the *methodology* of PTI: quantum theory simply refers to an underlying reality which includes advanced states. This is the simplest explanation of the form of the theory, including the Born Rule. Adding richness to the ontology is evidence of fruitfulness of the interpretation (just as the atomic hypothesis was a fruitful one), rather than an unwarranted complication.

9

Epilogue: more than meets the eye

9.1 The hidden origins of temporal asymmetry[1]

The model I've presented here is obviously time-asymmetric, since the fabric of spacetime grows in a particular direction with respect to the time parameter. Yet it is based on an underlying time-symmetric theory whose dynamics involves the propagation of both retarded and advanced fields, as in the Wheeler–Feynman/Dirac theories. If the underlying theory is time-symmetric, the only way we can arrive at the time-asymmetry of our experience, and of our model, is by reference to asymmetric boundary conditions (BC). As noted in Chapter 6, an important source of this asymmetry is that creation must precede destruction.

9.1.1 Time-isotropy vs. time-reversibility

To more fully understand this model, it is helpful to distinguish two features of symmetry that are often conflated: (1) the isotropy (or bidirectionality) of a process with respect to a given coordinate; and (2) the irreversibility of certain processes with respect to that same coordinate. Feature (1) is violated with respect to time in our empirical experience via the observation that temporal events constitute an ordered sequence that proceeds, like a set of movie frames, in only one direction (i.e., unidirectionally as the index t increases or decreases). This is qualitatively different from the case of spatial events which (in one spatial dimension x) can be bidirectional or (in three spatial directions) can be omnidirectional, like an expanding spherical wave. In Minkowski 4-dimensional (3+1) space, an empirically observable hyperspherical wave expands isotropically (in all directions) with respect to space, but only unidirectionally with respect to time.

[1] This section is based on material in Kastner (2011c).

It might be countered that we can only see what is in front of us in space at any given moment, and that we might think of this as making space and time equivalent in that sense. However, we can easily turn around and look in the back of us to confirm the existence of processes that are extended bidirectionally with respect to x, y, z in space. If one wants to eliminate the necessity for the head turn, we could set up a camera on our heads to record the scene behind us and feed it into a screen in front of us, so that we can see both what's in front of us and what's at the back of us. There is no way to experience a temporal analog of these procedures. While relativity tells us that space and time share a closer kinship than once thought, it must be kept in mind that the temporal coordinate is imaginary (ict) while the spatial coordinates x, y, z are real. This mathematical fact indicates that time and space are fundamentally distinct, and that time cannot be fully "spatialized."

Moreover, it is well known that in quantum mechanics one cannot construct a time operator, whereas one can construct a space (position) operator. The lack of a time operator is attributable to the fact that energy, the quantity conjugate to time, cannot be unbounded below (i.e., cannot be infinitely negative) for empirical events; in contrast, there is no such restriction on the quantity conjugate to position, i.e., momentum. So quantum mechanics also instructs us that time and space have a distinct status, and this is reflected in the Minkowski metric which assigns the temporal coordinate an imaginary status. (This fact constitutes another deep connection between quantum theory and relativity.)

Feature (2), time-reversal symmetry, is the observation, familiar from thermodynamics, that if we "run the movie backwards" most macroscopic processes (such as cream mixing into coffee) look physically unreasonable; while similar time-reversals of microscopic processes (such as small numbers of air molecules in a box) look reasonable. The former processes are termed "irreversible" and the latter "reversible." A fully isotropic process would be trivially time-reversal symmetric (it would be exactly the same under reversal); a merely reversible process, under time-reversal, would constitute a mirror-image process with (roughly) equal probability as the initial one. In contrast, an irreversible process, under time-reversal, would constitute a highly improbable process. But it is important to keep in mind that there is a distinction between (2) the observation that "running the movie backwards" can produce unrealistic phenomena and (1) the observation that *there is a "movie" in the first place* (events unfold anisotropically with respect to time). It is to (1) that we next direct our attention, since it is more fundamental.

9.1.2 Methodological and historical considerations

The Wheeler–Feynman and Davies theories, which provide an established theoretical basis for TI and PTI, are in fact *isotropic* theories with respect to field propagation; they do not just describe temporally reversible processes as in (2) above. These are conventionally called "time-symmetric" theories, so let's revert to that usage, keeping in mind that "time-symmetric" here really means "time-isotropic," not just "symmetric with respect to time reversal," which is a weaker condition.

One might wonder why we should entertain a time-symmetric theory, which seems counterintuitive in view of the points made above. One motivation for doing so is that our unidirectional temporal experience can be seen as evidence of a broken symmetry. Symmetries are broken by way of boundary conditions, such as those arising from the constraints on the milk droplet (recall Chapter 4) as it hits the pool, even if they don't determine the ultimate position of the milk droplet coronet. Just as we don't apply any restrictive boundary condition to the droplet *until* it hits the pool, the most general (and thereby most powerful) approach is to refrain from imposing boundary conditions on the theory until the underlying law is confronted with contingent features of our universe. The methodological advantage of this approach is that it avoids imposing possibly ad hoc explanations and conditions which may not actually hold in our universe; instead, it allows the theory itself to tell us what is required for the contingent asymmetries that we experience.

It should also be noted that specific boundary conditions must obtain in order for energy to be propagated by way of fields such as those obeying the wave equations of quantum field theory. That is, if one assumes a point source for the field (this corresponds to the inhomogeneous field equation), the basic (Green's function) solution to a wave equation is singular (undefined) for real energies. One cannot obtain a solution without analytic continuation (extension into the complex plane) of the energy coordinate, and choice of a contour of integration, the latter being determined by the relevant boundary conditions. This suggests that *fields associated with sources cannot actually propagate energy* unless specific boundary conditions (corresponding to a choice of integration contour in the complex plane) exist. In some sense, energy is only propagated due to the possibility of complex energies – corresponding to virtual particles or "propagators" in relativistic field theories. This subtlety concerning the ontology of energy propagation is routinely overlooked in most discussions of Green's functions and their various forms, and again suggests that sub-empirical processes, corresponding to the complex energy values, are in play in any propagation of actual (real) energy.

One of the propagators alluded to above is the "retarded" propagator which allows for field propagation only in the forward time direction (future). This solution is the one used in classical electromagnetic theory; the advanced solution, which allows for field propagation into the past, is simply dropped as "unphysical." However, it turns out that this approach does not account for the loss of energy ("radiative damping") by a radiating particle, and an ad hoc additional free field must be assumed (see below).

Wheeler and Feynman (WF) decided to explore a time-symmetric approach to field propagation because they were not satisfied with the standard method of dealing with radiative damping. Dirac (1938) had proposed that damping can be explained by a free field (that is, a field not attributed to any source) in addition to the assumed fully retarded (unidirectional, positive *t* direction) propagation due to the charge. While this seemed to account for the observed energy loss, WF were dissatisfied with its ad hoc character. They proposed instead that the basic propagation due to the charge was time-symmetric; the time-symmetric propagator is simply the sum of half the retarded and half the advanced propagators. WF proposed that other charges responded to that time-symmetric field by emitting their time-symmetric field out of phase with the stimulating field. If the universe is a "light-tight box" – that is, if propagation from sources is not allowed into the infinite future – the response of the absorbing particles turns out to provide, at the location of the emitting charge, the apparent "free field" needed to account for loss of energy by the radiating charge, as well as cancellation of the retarded field beyond the absorbers (which is why they absorb), and of advanced propagation (of positive energy into the past) due to the radiating charge. Thus the asymmetric boundary condition of no infinite future propagation provides for the apparent time-asymmetry of radiation (i.e., its unidirectionality), as well as a natural (not ad hoc) explanation for the absorption of energy and of radiative damping.

The WF theory thus describes radiation of energy as a direct interaction between sources (sinks), and the emitting particle is taken as not interacting at all with its own emitted field (the latter process is commonly referred to as "self-interaction"). Such theories are called "direct action" (DA) theories. However, it later became evident that some form of self-interaction was needed to account for certain relativistic effects (such as the Lamb shift).[2] Davies introduced this feature into his quantum relativistic extension of the Wheeler–Feynman theory (Davies, 1971, 1972), which was discussed in Chapter 6.

It remains a matter for further investigation as to whether DA-type theories are empirically equivalent to standard quantum field theories for our universe. However, to date there is no conclusive evidence that DA theories are not empirically equivalent, and there is much to recommend them in methodological terms, as argued above.[3] Some researchers dislike DA theories because they are generally

[2] Lamb and Retherford (1947).

[3] If the time-symmetric process of TI and PTI is considered as a direct action theory as proposed by WF and Davies, then perfect correspondence with the observed asymmetry of energy transfer would appear to depend on either an opaque future universe or a perfectly reflecting past universe (see, e.g., Cramer, 1983, which discusses a perfectly reflecting boundary condition). This is a matter for further cosmological research. It is often asserted that the universe is not opaque as required for a direct action theory, but this does not rule out the perfectly reflecting

impractical for doing computations, since they depend explicitly on the actual boundary conditions of the generating sources and sinks (for which the technical term is "currents") both in the future and in the past, and the directly acting field does not have independent degrees of freedom of its own that can be quantized (i.e., treated as harmonic oscillators with discrete states of excitation). In contrast, standard field theories use quantized fields as independent entities and therefore do not need to refer explicitly to the currents that generate them. While standard quantum field theories are therefore much better computational tools, that pragmatic fact does not rule out the idea that nature actually uses direct action in the universe, of which we can only study a small portion in any given computation.

9.1.3 Boundary conditions and the arrow of time

As noted above, assuming only retarded field propagation does not allow for radiative damping; an ad hoc free field must be imposed. A more natural and general approach takes the underlying propagation as time-symmetric (isotropic) and seeks to discover what actual boundary conditions must exist in order for energy to be transferred from one place to another in accordance with empirical observation. As discussed above, this turns out to be the condition that the universe is a "light-tight box"; i.e., fields do not propagate to infinity.[4] Another way of understanding this condition is that there are many microscopic absorbers for each microscopic emitter, since any emitted field must be absorbed somewhere (see, e.g., Davies, 1972, p. 1046).[5]

It has been observed in the past (e.g., Ritz, 1909 as quoted in Zeh, 1989, p. 13) that one could relate the apparent asymmetry of electromagnetic radiation to the thermodynamic "arrow" – i.e., the preponderance of irreversible physical processes (such as the mixing of coffee and milk discussed above). However, those approaches simply omitted the advanced electromagnetic wave solution *a priori* and assumed that the resulting asymmetry of the retarded solution alone could be extended to the general thermodynamic asymmetry. This was problematic because

boundary condition, which achieves the same empirical retarded radiation result. Moreover, the jury is still out on the ultimate structure of the universe. I believe it would be a mistake to rule out any particular interpretation based on what appears to be the case in current cosmological studies. Recall that Einstein referred to the "cosmological constant" as his "biggest blunder," but it has since been thoroughly rehabilitated as a crucial component of current cosmological theory.

[4] As noted earlier, if the universe is not a "light-tight box," then one can have radiation (energy) propagating to infinity by way of the perfectly reflecting t=0 boundary condition. The latter is just another kind of asymmetric BC.

[5] It should be noted here that many extant discussions of this issue (e.g., Callender, 2002) assume that retarded fields only describe emission and advanced solutions only describe absorption of radiation – i.e., that the latter describe phenomena surrounding radiation sinks. However, this is one proposal among many, and does not address radiative damping which remains unexplained except for the ad hoc free field of Dirac. See also Price (1996).

the thermodynamic asymmetry applies to neutral particles as well, and would therefore seem to have nothing to do with electromagnetic fields.

If, in contrast to the traditional *a priori* rejection of advanced solutions, we suppose that the underlying laws are truly time-symmetric, then the apparent asymmetry of radiation (i.e., retarded only) is due to the asymmetry of boundary conditions. This approach to electrodynamics can readily be extended, via TI, to neutral particles, since TI applies to all quantum fields, including neutral ones. While thermodynamic irreversibility is usually thought of as applying to macroscopic (classical) systems, it should be kept in mind that the quantum level is the more fundamental one which must underlie all classical phenomena. The transactional interpretation of quantum theory tells us that all transfers of energy (and other conserved physical quantities) take place due to the interaction of the emitted field with absorbers conforming to the boundary conditions needed for cancellation of residual advanced and retarded fields – where the latter are *all* quantum fields, not just charged ones.

The picture that emerges is the following: symmetrical physical laws describe theoretical potentialities, not actualities. In order to have actual events in an actual world, the symmetries of those laws must be broken by the imposition of constraints in the form of boundary conditions. Such boundary conditions may not always specify *which* actual event or form will exist – often that specific event will arise from spontaneous symmetry breaking – but they precipitate that actuality. In the case of energy propagation, the boundary condition is the preponderance of microscopic absorbers compared to emitters (or the perfectly reflecting $t = 0$ boundary condition discussed in Cramer, 1983), together with the principle that emission has ontological priority over absorption and the absorber radiates exactly out of phase with the stimulating emitted field.

Thus we can see the "arrow of time" as a result of symmetry breaking of physical laws governing energy propagation. The same boundary conditions necessary for propagation of energy in an underlying time-symmetric theory may serve to explain thermodynamic irreversibility when that theory is extended to the quantum domain which underlies all macroscopic processes. Thus the most economical and natural explanation of both aspects of the "arrow of time" is that basic physical laws are symmetrical with respect to both space and time but describe only potentialities, and that actual events and processes arise because of symmetry breaking.

Since the direction of positive energy transfer dictates the direction of change, and time is precisely the domain of change (or at least the construct we use to record our experience of change), it is the broken symmetry with respect to energy propagation that establishes the unidirectionality of time.[6] The reason for the

[6] Recall also that energy is conjugate to time in quantum mechanics.

"arrow of time" is that the symmetry of physical law must be broken for actual events to occur: "the actual breaks the symmetry of the potential."

9.2 Concluding remarks

This book has endeavored to present an interpretation of quantum theory that takes into account as much as possible of the mathematical formalism of the theory. My approach assumes that quantum theory is "complete" in the sense that it accurately and completely describes the quantum domain. That is, I do not believe that the theory needs to be extended or to be either ontologically or mathematically modified, as is assumed in other approaches such as that of the Bohmian theory (which adds corpuscles not directly referred to by the mathematics) and GRW theories (which add an explicit non-linear component); nor do I believe that quantum theory is about the knowledge or ignorance of an observer.

The interpretive challenge of quantum theory is often presented in terms of the measurement problem: i.e., that the formalism itself does not specify that only one outcome happens, nor does it explain why or how that particular outcome happens. This is the context in which it is often asserted that the theory is incomplete or inexact and is therefore in need of alteration in some way. However, as we have seen in Chapter 4, there are situations in classical physics and in the standard model of elementary particle theory in which a very similar situation occurs: the theoretical formalism describing a situation specifies a set of solutions, but does not specify which one will occur nor explain how it occurs. This is described by the authors of *Fearful Symmetry* as an apparent violation of Curie's principle. They address the situation by suggesting that Curie's principle should be understood in an "extended" sense: the symmetry is still there, but it is hidden (and ultimately broken). Note that in the above cases, the situation is not considered to pose an intractable "measurement problem" nor to require a "many worlds" interpretation; i.e., that all outcomes must be assumed to occur. In the classical case, one can point to some tiny quantum fluctuation as Curie's "asymmetric cause" of the actualized outcome; it is not possible to point to any specific external asymmetric cause in the case of spontaneous symmetry breaking in the Standard Model. Such an agent would again have to be some sort of quantum fluctuation, unless we want to postulate some further subquantum domain. If we are going to help ourselves to a quantum (or subquantum) fluctuation as the agent of actualization of one of a set of possible solutions in these cases, then why can't we do the same for standard quantum theory?

If we adopt the approach that quantum theory tells us about many possibilities arising from interactions between offer waves and confirmation waves, then we gain a clearly defined set of possible outcomes[7] missing in the standard account, which disregards the real physical process of absorption. Recall that one component of the measurement problem is the amplification of the quantum state through interactions with the measuring apparatus, the first observer, the second observer, etc., with no means of deciding when the measurement has been completed. The designation of a stage at which the measurement is "completed" is referred to as the "Heisenberg cut," which is notoriously arbitrary. The arbitrariness is removed once we notice that the original offer wave inevitably encounters one or more absorbers that generate confirmations in response to the offer. It is at that point that a set of incipient transactions is established. This set of incipient transactions is perfectly analogous to the set of solutions appearing in standard cases of symmetry breaking, except for the additional feature that the set of incipient transactions may have unequal weights. This interesting additional feature can be interpreted as indicating that some possibilities are more potent than others, but that they are not guaranteed; that their actualization is uncertain, in the same qualitative way as any specific outcome is uncertain in a case of spontaneous symmetry breaking.

Recall again the splashing milk droplet of Figure 4.5 (I reproduce it here as Figure 9.1 for convenience). This iconic photograph records a process of spontaneous symmetry breaking, with a particular outcome having been actualized in that the points of the crown occur at a set of particular locations as opposed to some other (rotated) set of locations. The standard discussion of this process goes like this: the spherically symmetric milk droplet hits the circularly symmetric reservoir of milk, causing a circular ring of milk to rise. As it rises, it becomes thinner and thinner. At some dynamical critical point, the ring's shape cannot be supported by its material components and it must make a "decision" whose form is dictated by the 24-pointed polygon (the geometry underlying the "crown") but whose actual orientation is not. In the photograph we witness a profound mystery: *something specific happens, even though there is no specific (deterministic) "reason" for it.* In a perfectly analogous manner, when an offer wave is split between two or more competing absorbers, both of which respond with confirmation waves, a dynamical "critical point" is reached in which some decision must be made. If we have to point to a random quantum fluctuation that "pushes" the system to one choice or the other, that is not qualitatively different from what we do in the milk droplet case (or the Higgs mechanism case). The only difference is that, in the quantum case, some outcomes can be more likely than others.

Thus the natural interpretation of quantum theory is that it tells us about a vast unseen network of *potential* events underlying our visible, actual, "tip of the

[7] That is, a specific basis in which the weighted set of projection operators of the system's density matrix possesses a clear physical referent.

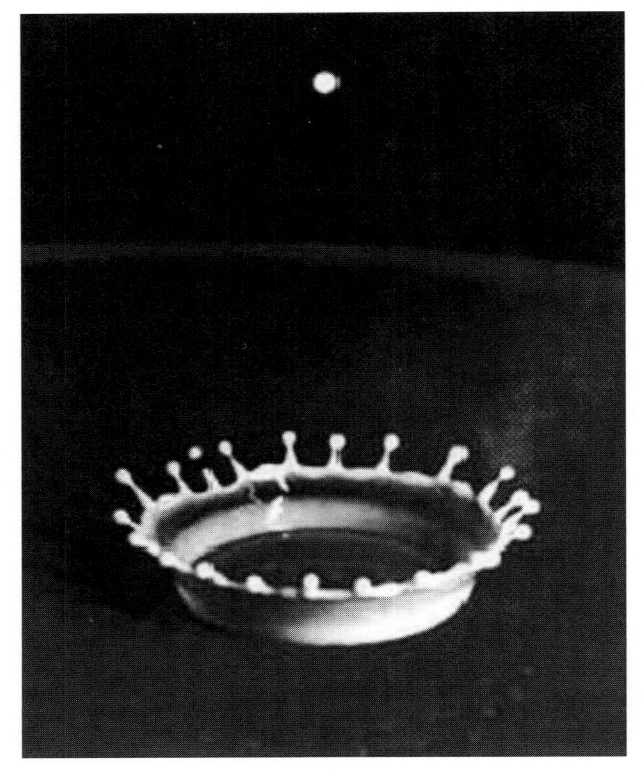

Figure 9.1 The iconic symmetry-breaking milk droplet: the back-reaction of the reservoir must be taken into account.
Source: Harold E. Edgerton, Milk-Drop Coronet, 1957. © 2010 Massachusetts Institute of Technology. Courtesy of MIT Museum

iceberg" world. It tells us that there are far more potential events than can be actualized; and that actualization occurs by way of confirmation waves from absorbers and a process of spontaneous symmetry breaking. Confirmation waves are analogous to the reservoir of milk that receives the falling milk droplet. It is the interaction between the droplet and the reservoir which precipitates the set of possible outcomes and the actualization of one of them. In the same way, it is the interaction between an offer wave and a set of absorbers which precipitates the set of possible outcomes (incipient transactions) and the actualization of one of those.

Roughly speaking, we can think of the usual, unsuccessful ways of trying to interpret quantum theory as a process of trying to interpret a falling milk droplet that never encounters anything, or a droplet encountering other objects but the back-reaction of those objects upon the droplet is not being taken into account. It is the omission of the back-reaction which gives rise to the notorious intractability of the measurement problem. Once that reaction (i.e., the confirmation) is included, we are back to a form of spontaneous symmetry breaking, and nature makes its choice: the actual arises from the potential, and we experience the tip of the iceberg that rests upon a vast, unseen, submerged body of potentiality in an even vaster "ocean" of possibility.

Appendix A

Details of transactions in polarizer-type experiments

The presentation of polarizer experiments in Cramer (1986) is somewhat compact and omits explicit discussion of the component of $|\mathrm{S}\rangle$ blocked by the initial polarizer, which, for example, transmits horizontally polarized light in the state $|\mathrm{H}\rangle$. It is noted therein on p. 657 that the source emits light in an arbitrary state of polarization and that only the component $|\mathrm{H}\rangle$ is transmitted.

Note 19 comments that the non-transmitted component can't form a transaction beyond the filter. From then on, that component is neglected and the transmitted portion treated as normalized to unity. Here I fill in some of those details to see how there is no ambiguity or arbitrariness in the transaction amplitudes.

Recall that the relevant polarization states are related by

$$|\mathrm{R}\rangle = 1/\sqrt{2}(|\mathrm{H}\rangle + i|\mathrm{V}\rangle) \tag{A.1a}$$

$$|\mathrm{H}\rangle = 1/\sqrt{2}(|\mathrm{R}\rangle + i|\mathrm{L}\rangle) \tag{A.1b}$$

where $|\mathrm{V}\rangle$ is vertically polarized light, orthogonal to $|\mathrm{H}\rangle$, and $|\mathrm{R}\rangle$ and $|\mathrm{L}\rangle$ are right- and left-circularly polarized light, respectively.

Let the source state $|\mathrm{S}\rangle$, an arbitrary state of polarization, be expressed in terms of the transmitted and absorbed components, $|\mathrm{H}\rangle$ and its complement $|\mathrm{V}\rangle$:

$$|\mathrm{S}\rangle = a|\mathrm{H}\rangle + b|\mathrm{V}\rangle \tag{A.2}$$

where $a^*a + b^*b = 1$.

The following discussion refers to Figure A.1. The top set of arrows represents the process in which a horizontally polarized OW component $|\mathrm{H}\rangle$ proceeds to a filter for right-circularly polarized light, is attenuated by an amount $1/\sqrt{2}$, is absorbed, and then a CW proportional to $1/\sqrt{2}\langle R|$ emitted. This component continues through the H filter where it is further attenuated by another factor of $1/\sqrt{2}$ for a final amplitude proportional to ½. When we neglect the initial attenuation of the beam by its passage through H, the constant of proportionality can be taken as unity. However, in reality the initial OW $|S\rangle$ is attenuated by a factor of a as it proceeds through the H filter. This factor goes "along for the ride" in the above computation, turning into its complex conjugate a^* when the CW is emitted, with this additional step: as the returning CW, in the state $a^*/2\langle H|$, reaches the source in state $|S\rangle$, it is projected onto state $|S\rangle$ and picks up another factor of a (from $\langle H|S\rangle = a$, see (A.2)). Thus the *truly* final CW amplitude for this component at the emitter location will be $a^*a/2$. This is necessary in order for the CW to interact with the source in accordance with conservation laws, and for cancellation of the remaining CW proceeding into the past, i.e., for times $t < T$ with respect to the emission time T. The latter feature will be discussed further below.

The lower set of arrows represents the component blocked by the filter, corresponding to the state $|V\rangle$. Its amplitude at the filter is b; thus the CW returning to the source is $b^*\langle V|$. As in the above process, this component's projection onto the source state $|S\rangle$ is given by $\langle V|S\rangle = b$. Thus the final amplitude of this component back at the source is b^*b.

Not explicitly shown in the diagram, but also needing to be taken into account, is the OW/CW combination corresponding to the beam blocked by the R filter. The final amplitude of this component will be $a^*a/2$,in view of the relation (A.1b) which tells us that equal amounts of the $|H\rangle$ beam are transmitted and blocked by the R filter.

So we have three possible transactions in this experiment: (1) the photon is blocked by the first filter, with a probability of b^*b; (2) the photon is blocked by the second filter, with a probability of $a^*a/2$;(3) the photon is transmitted through both filters and detected at D, with a probability of $a^*a/2$. The probabilities for these three transactions, each equal to the final amplitude of the returning confirmation wave at the point where it is back in the state $\langle S|$, sum to unity.

Note that if we follow the CW components into the past, their amplitudes will sum to unity, giving rise to a unit advanced wave $\langle S|_F$ where the subscript indicates

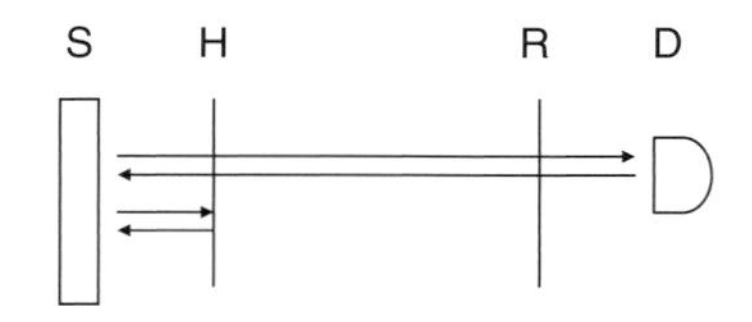

Figure A.1 Possible transactions in a polarizer experiment.

they originated from a detection point in the future of S. But they will be perfectly out of phase with the original advanced wave $\langle S|_S$ emitted by the source, and as in the discussion in Cramer (1986, pp. 661–3), the two advanced waves will destructively interfere, leaving only the possible transactions between the source and the detection points.

The discussion in Cramer (1986) considers only the transmitted component $|H\rangle$ and therefore implicitly normalizes the OW by a factor of a^*a. But there is no arbitrariness or inconsistency in the general procedure of calculating the amplitudes of the transactions. The final amplitude of the CW back at the location of the emitter must take into account the projection of the CW onto the original state of the source, and this is what gives rise to the symmetrical form, $\Psi^*\Psi$, of the Born Rule.

Appendix B

Feynman path amplitude

It was noted in Chapter 4 that one can define the Feynman "amplitude of a path" for a quantum particle in the context of the Feynman picture. However, this does not correspond to a well-behaved probability in the absence of a sequence of actualized transactions defining the associated trajectory.

Using the Feynman sum-over-paths method, one obtains the probability to go from point A to B by summing over all possible "paths" from A to B to get an amplitude, and then squaring that amplitude. Let us simplify this as in the first chapter of Feynman and Hibbs (1965), wherein the space between A and B is subdivided by a finite number of intermediate stages, say C, D, and E.

One first obtains the amplitude Amp(AC) to go from A to C, then similarly from C to D, from D to E, and finally from E to B. The total amplitude to go from A to B by way of C, D, and E is the product of these amplitudes: $\mathrm{Amp(AB)} = \mathrm{Amp(AC)} \times \mathrm{Amp(CD)} \times \mathrm{Amp(DE)} \times \mathrm{Amp(EB)}$.

Now, if there is no other way to get from A to B, the associated probability is the absolute square of Amp(AB), i.e.:

$$\begin{aligned}\mathrm{Prob(AB)} = \{&\mathrm{Amp(AC)} \times \mathrm{Amp(CD)} \times \mathrm{Amp(DE)} \times \mathrm{Amp(EB)}\} \\ &\times \{\mathrm{Amp}^*(\mathrm{AC}) \times \mathrm{Amp}^*(\mathrm{CD}) \times \mathrm{Amp}^*(\mathrm{DE}) \times \mathrm{Amp}^*(\mathrm{EB})\}\end{aligned} \tag{B.1}$$

Mathematically, we can just rearrange this to get:

$$\begin{aligned}\mathrm{Prob(AB)} = \{&\mathrm{Amp(AC)} \times \mathrm{Amp}^*(\mathrm{AC})\} \times \{\mathrm{Amp(CD)} \times \mathrm{Amp}^*(\mathrm{CD})\} \\ &\times \{\mathrm{Amp(DE)} \times \mathrm{Amp}^*(\mathrm{DE})\} \times \{\mathrm{Amp(EB)} \times \mathrm{Amp}^*(\mathrm{EB})\}\end{aligned} \tag{B.2}$$

If there *are*, however, other ways to get from A to B, we still might be tempted to assume that we can define a "probability" to go between each of the intermediate stages, i.e.:

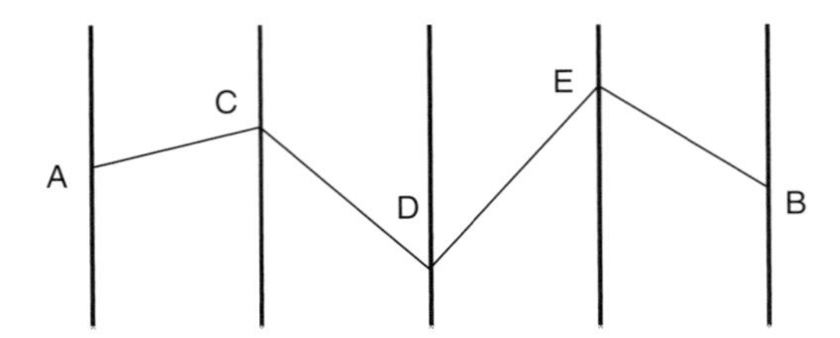

Figure B.1 A path from A to B.

$$\text{Prob}(AB) = \text{Prob}(AC) \times \text{Prob}(CD) \times \text{Prob}(DE) \times \text{Prob}(EB) \quad \text{(B.3)}$$

implying that there exists a physically meaningful "probability for a particular path" such as $A \rightarrow C \rightarrow D \rightarrow E \rightarrow B$. However, under TI, the only reason you multiply an amplitude by its complex conjugate is because a confirmation occurs. If there is no absorber at the points C, D, or E, there is no independent complex conjugate factor such as Amp*(AC), etc.

Moreover, such "partial amplitudes" as Amp(AC) do not correspond to well-behaved probabilities. This is well known in the context of the two-slit experiment, where the amplitudes to go from a source to a point x on a final screen by way of slits A or B do not correspond to probabilities that are additive (recall Section 3.3.2). In this sense, there is an "amplitude to go from the source to the slit by way of slit A (or slit B)," but that does not correspond to a meaningful probability that a particle actually went one way or the other, *unless* there are absorbers at the intermediate points (i.e., unless we have a detector to see "which slit the particle went through"). This *is* the case in a bubble chamber, so one can define a "path" for a quantum particle in a bubble chamber due to the interaction of the OW with absorbers in the chamber. We should not, however, let this lead us to think that outside the bubble chamber, the particle pursues a particular spacetime trajectory. The reason that you must add the amplitudes for all possible ways to go from A to B is because the quantum (i.e., offer wave) *does* pursue all those possible ways (it does this in PST, not in spacetime).

Appendix C

Berkovitz contingent absorber experiment

Berkovitz (2008) presents a variation on the Maudlin experiment which he terms "Experiment X" (see Figure C.1). This is an EPR-type experiment involving the usual "singlet" spin state

$$|\varphi\rangle = \frac{1}{\sqrt{2}}[|\uparrow,\downarrow\rangle - |\downarrow,\uparrow\rangle] \tag{C.1}$$

The spin-measuring apparatus on the right, R, is within the past light cone of the spin-measuring apparatus on the left, L. R is fixed to measure spin along direction r in all runs of the experiment. The setting of L is made contingent on the outcome at R as follows. If the outcome observed at R is "down," then L measures along the same direction $l = r$. If the outcome observed at R is "up," then L measures along a different direction $l^* \neq r$. Berkovitz argues that the application of TI to the experiment gives rise to a set of causal loops which prevent the prediction of probabilities of outcomes. He claims that the existence of the fixed incipient transaction between the emitter and the detector R is insufficient to specify these probabilities, and that the approach in Kastner (2006) is therefore not sufficient. However, one need only modify that account in a way appropriate to the experiment to resolve this issue; thus it does not constitute a new challenge above and beyond the Maudlin experiment. The modified account consists in observing that there are *two* sets of transactional processes at play in this experiment, in contrast to the one set (corresponding to the OW components to the right and left) in the Maudlin example.

For convenience, let us suppose that the "default" setting of L is l; in order to measure l^*, a (subluminal) signal must be sent from R to change L's setting. This will only happen when the R outcome is "up." (Note that this is arbitrary; we could just as well use the opposite convention, having the default be l^* with the changing signal for measurement of l being sent when the R outcome is "down.") There are actually two absorbers for R and L, one that absorbs "up," R+(L+), and the other that

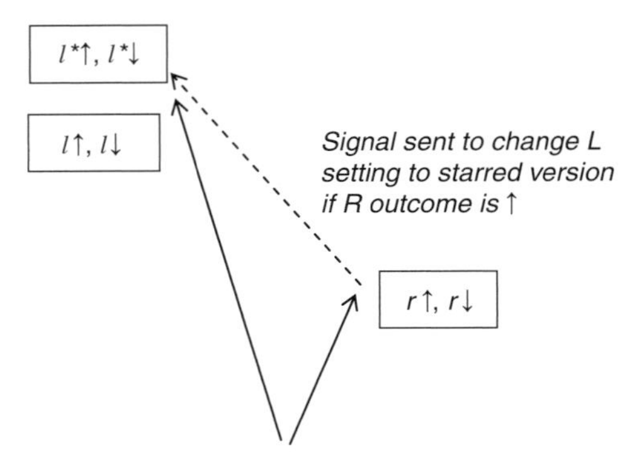

Figure C.1 Experiment X.

absorbs "down," R–(L–). R+,– can only generate single-particle CWs corresponding to the right-hand particle and L+,– can only generate single-particle CWs corresponding to the left-hand particle.

The two-particle singlet-state offer wave can be written, in the "r" basis ($r = l$), as

$$|\varphi\rangle = \frac{1}{\sqrt{2}}[|r\uparrow\rangle\otimes|l\downarrow\rangle + i|r\downarrow\rangle\otimes i|l\uparrow\rangle] \tag{C.2}$$

where I make the direct product between particle subspaces explicit for expositional purposes below. If we were to add "labels" to the particles, the r labels particle "1" and the l labels particle "2" (I leave off the labels since this is indicated by their order, but this issue of labeling is relevant to what follows). The r, l basis is arbitrary; the basis of any incipient transactions will be determined by what types of absorbers are present, as will be seen below.

Implicit in any detection of a single particle from an emitted two-particle state is a reduction of the spatial component of the total state to one for a single particle. Recall that the spatial component $|\psi\rangle$of the total antisymmetric two-electron state (for the spin singlet), $|\Psi\rangle = |\psi\rangle|\varphi\rangle$, is a symmetric superposition of particle labels, i.e.

$$|\psi\rangle = \frac{1}{\sqrt{2}}[|\mathrm{R}(1)\rangle\otimes|\mathrm{L}(2)\rangle + |\mathrm{L}(1)\rangle\otimes|\mathrm{R}(2)\rangle] \tag{C.3}$$

where $|\mathrm{R}\rangle$ and $|\mathrm{L}\rangle$ are spatial states propagating to R and L, respectively. The indistinguishability of the particles means that R+,– does not care whether it is detecting "1" or "2" – such labels contain no genuine information for indistinguishable objects – but for consistency the single-particle CWs must correlate labels for the spatial and spin states. Thus, if we take R to be detecting particle "1," it cannot know anything about the spin of particle "2" heading to the left (and which, as noted above, is designated by the second ket labeled l in each term of (C.2)). Thus the spin states for the left-hand particle are invisible to it and it cannot respond to them. The random "choice" of one of the two particles (whether it is "1" or "2" is arbitrary; here we choose "1") by R+,− leaves for the total right-hand CW (i.e., the CWs from R+ and R−):

$$\langle\Psi_{\mathrm{R}}| = \langle\psi_{\mathrm{R}}|\,\langle\varphi_{\mathrm{R}}| = \langle \mathrm{R}(1)|\left[\frac{1}{\sqrt{2}}\langle r\uparrow| - \frac{i}{\sqrt{2}}\langle r\downarrow|\right] \tag{C.4}$$

In (C.4), the term $\langle \mathrm{R}(1)|\left[\frac{1}{\sqrt{2}}\langle r\uparrow|\right]$ is the space/spin confirmation wave from R+ and $\langle \mathrm{R}(1)|\left[\frac{-i}{\sqrt{2}}\langle r\downarrow|\right]$ is the space/spin confirmation wave from R–. (The minus sign appears because of complex conjugation for the advanced state; see also below.) Thus, while there is no pure state single-particle offer wave, each particle's confirmation wave must have the form of a single-particle brac and must reflect the properties of the absorber (see note 1).

Recall that the amplitude of a CW at the locus of the absorber is the complex conjugate of the amplitude of the offer wave component absorbed by it. Thus, absorption by a detector "A" of an OW component of state $|a\rangle$ with an amplitude of a, i.e., $a|a\rangle$, results in a CW of $a^*\langle a|$. Their overlap (incipient transaction) is described by the outer product $a^*a|a\rangle\langle a|$ (revealing the Born Rule). According to the analysis in Cramer (1986), a^*a is the amplitude of the relevant CW component as it reaches the emitter of its prompting OW, but it is convenient to think of the entire expression $a^*a|a\rangle\langle a|$ as applying to the superposition of a particular set of matching OW and CW components; i.e., an incipient transaction. Thus, the establishment of the incipient transactions for the system based on the default L setting is represented by the diagonal terms of the outer product of the total emitted state and the total right- and left-hand CWs: all off-diagonal terms vanish, because an absorber for a particular state cannot absorb its complement (i.e., the state orthogonal to the one absorbed).[1] Also, total system spin CWs of the form $\langle r\uparrow|\otimes\langle l\uparrow|, \langle r\downarrow|\otimes\langle l\downarrow|$ cannot participate in any viable (probability greater than zero) transactions by conservation laws (in this case, conservation of angular momentum).

[1] This is a standard feature of detections; an example is a horizontal polarizing filter which absorbs vertically polarized photons but not horizontally polarized ones, which it passes unaffected. Thus a horizontal polarizing filter, which functions as an absorber/detector for vertically polarized photon offer waves $|\mathrm{V}\rangle$, generates only vertically polarized confirmation waves $|\mathrm{V}\rangle$. There is no physical situation that can result in incipient transactions of the form $|\alpha\rangle\langle\beta|$ where α and β are orthogonal. If this were so, we would have situations such as horizontally polarized photons being blocked/detected by a horizontal polarizing filter (i.e., the incipient transaction represented by $|\mathrm{H}\rangle\langle\mathrm{V}|$ would have a finite possibility of being an actualized transaction).

Thus, we have for the total system's set of incipient transactions (suppressing the spatial component):

$$\left\{|\varphi\rangle\left[\frac{1}{\sqrt{2}}\langle r\uparrow|-\frac{i}{\sqrt{2}}\langle r\downarrow|\right]\otimes\left[\frac{1}{\sqrt{2}}\langle l\downarrow|-\frac{i}{\sqrt{2}}\langle l\uparrow|\right]\right\}_{\text{diag}}=$$

$$\left\{\frac{1}{2}[|r\uparrow\rangle\otimes|l\downarrow\rangle-|r\downarrow\rangle\otimes|l\uparrow\rangle][\langle r\uparrow|\otimes\langle l\downarrow|-\langle r\downarrow|\otimes\langle l\uparrow|+(c.t.)]\right\}_{\text{diag}}=$$

$$\frac{1}{2}|r\uparrow\rangle\otimes|l\downarrow\rangle\langle r\uparrow|\otimes\langle l\downarrow|+\frac{1}{2}|r\downarrow\rangle\otimes|l\uparrow\rangle\langle r\downarrow|\otimes\langle l\uparrow| \quad \text{(C.5)}$$

where "(c.t.)" stands for cross-terms such as $\langle r\uparrow|\otimes\langle l\uparrow|$ which violate conservation laws as discussed above. (The total system brac in the first term of (C.5) may look unfamiliar; what it represents is simply all CW components present. As discussed previously, each detector can only respond with single-particle confirmation waves.)

We see again here the transactional basis for the von Neumann "weak projection" postulate. Each term in (C.5) represents a weighted incipient transaction for the whole system. The left-hand particle's offer wave state is clearly defined based on an actualized R-transaction, since R is in the past light cone of L.

According to the experimental setup, if the outcome of R is "down," that means a transaction has occurred in which apparatus L will be unaltered and will measure $l=r$. Since the left-hand particle's offer wave is already $|l=\text{up}\rangle$ when it encounters the L+,− absorbers, this is a transaction with probability 1: it must occur with certainty. As noted earlier, if the default setting on the left is to measure l^*, it will be changed via a light signal to the l basis upon actualization of the transaction specifying $|l=\text{up}\rangle$ and the left-hand transaction will still be actualized with certainty.

On the other hand, if the outcome of R is "up," that means a transaction has occurred in which the left-hand particle must be found in the state "down along l," i.e., $|l=\text{down}\rangle$. A subluminal signal will be sent to apparatus L so that it will measure l^* (not equal to r). In this case, another transactional situation appears for the left-hand particle and the CW are the basis states of l^*. The probability of the outcome "up along l^*" is given by $|\langle l^*\uparrow||l\downarrow\rangle|^2$ and similarly for "down along l^*" (see Table C.1). Obtaining a determinate result for the left-hand particle when the R outcome is "up" will thus require actualization of a particular transaction from a further set of incipient transactions on the left.

Thus the basic solution to the predictivity challenge in terms of Berkovitz's picture (in which the OW and CW propagation is situated in spacetime) is as described in Kastner (2006): the unconditioned probabilities are defined with respect to the

Table C.1

L-setting	Probability of outcome "up"	Probability of outcome "down"
l	$\lvert\langle l = \text{up}\rvert\lvert l = \text{up}\rangle\rvert^2 = 1$	$\lvert\langle l = \text{down}\rvert\lvert l = \text{up}\rangle\rvert^2 = 0$
l*	$\lvert\langle l^* = \text{up}\rvert\lvert l = \text{down}\rangle\rvert^2$	$\lvert\langle l^* = \text{up}\rvert\lvert l = \text{down}\rangle\rvert^2$

weights of the incipient transaction(s) in the fixed portion of the experiment. In the present case, these incipient transactions constitute a time-symmetric "seed" or catalyst which determines which loop will occur: the loop in which, in Berkovitz's analysis, the "state of the emitter" is one in which the confirmations from L are in the *l* or *l** basis. It is not necessary that the same left-hand OW/CW incipient transactions exist in all runs of the experiment as long as there is a set sufficient to distinguish the two loops; and this is provided by the R incipient transactions whose outcomes (i.e., which one is actualized and which one is not) determine the type of confirmation waves that will exist on the left. So, as Berkovitz notes, the set of incipient R-transactions is not sufficient to specify all the probabilities, but there is an additional set of incipient transactions on the left in the event that the R-outcome is "up" that can provide them.

Thus TI unambiguously predicts the conditional probabilities of Table C.1, which can be measured as long-run frequencies, for the various L-settings. The probability of each L-setting is ½, based on the R-outcome available from the fixed portion of the experiment. From Table C.1 we can obtain the unconditional probabilities of the various combinations of outcomes for both particles; e.g., the probability that R detects "up" and L detects "down" is $\frac{1}{2}\lvert\langle l^* = \text{down}\rvert\lvert l = \text{down}\rangle\rvert^2$. Thus "Experiment X" is not qualitatively different from the Maudlin challenge involving an indeterminary of CW as addressed in Chapter 5.

References

Aharonov, Y. and Albert, D. (1981). "Can we make sense out of the measurement process in relativistic quantum mechanics?" *Physical Review D* **24**, 359–70.

Allison, W. W. M. and Cobb, J. H. (1980). *Annual Review of Nuclear and Particle Science* **30**, 253.

Anandan, J. (1997). "Classical and quantum physical geometry." In R. S. Cohen, M. Horne and J. Stachel (eds), *Potentiality, Entanglement and Passion-at-a-distance – Quantum Mechanical Studies for Abner Shimony*, Vol. **2**. Dordrecht: Kluwer, pp. 31–52. Preprint: http://arxiv.org/PS_cache/gr-qc/pdf/9712/9712015v1.pdf

Arndt, M. and Zeilinger, A. (2003). "Buckeyballs and the dual-slit experiment." In J. Al-Khalili (ed.), *Quantum: A Guide for the Perplexed*. London: Weidenfeld & Nicolson.

Auyang, S. (1995). *How is Quantum Field Theory Possible?* New York: Oxford.

Bacciagaluppi, G. and Crull, E. (2009). "Heisenberg (and Schrödinger, and Pauli) on hidden variables." http://philsci-archive.pitt.edu/archive/00004759/01/SHPMP_paper_07_10_09.pdf

Barbour, J. B. (1982). "Relational concepts of space and time." *British Journal for the Philosophy of Science* **33**, 251–74.

Barnum, H. (1990). "Dieks' realistic interpretation of quantum mechanics: a comment." Preprint: philsci-archive.pitt.edu/2649/1/dieks.pdf

Bell, J. S. (1990). "Against measurement." *Physics World*, August, 33–41.

Bennett, C. L. (1987). "Precausal quantum mechanics." *Physical Review A* **36**, 4139–48.

Berestetskii, V., Lifschitz, E. and Petaevskii, L. (1971). *Quantum Electrodynamics*. Landau and Lifshitz Course of Theoretical Physics, Vol 4. Amsterdam: Elsevier.

Berestetskii, V., Lifschitz, E. and Petaevskii, L. (2004). *Quantum Electrodynamics*, 2nd edn. Landau and Lifshitz Course of Theoretical Physics, Vol **4**. Amsterdam: Elsevier.

Berkovitz, J. (2002). "On causal loops in the quantum realm." In T. Placek and J. Butterfield (eds), *Proceedings of the NATO Advanced Research Workshop on Modality, Probability and Bell's Theorems*. Dordrecht: Kluwer, pp. 233–55.

Berkovitz, J. (2008). "On predictions in retro-causal interpretations of quantum mechanics." *Studies in History and Philosophy of Modern Physics* **39**, 709–35.

Bethe, H. (1930). "Zur Theorie des Durchgangs schneller Korpuskularstrahlen durch Materie." *Annalen der Physik* **397**, 325–400.

Bohr, A., Mottelson, B. and Ulfbeck, O. (2003). "The principle underlying quantum mechanics." *Foundations of Physics* **34**, 405–17.

Braddon-Mitchell, D. (2004). "How do we know it is now now?" *Analysis* **64**, 199–203.

Breitenbach, G. Schiller, S. and Mlynek, J. (1997). "Measurement of the quantum states of squeezed light." *Nature* **387**, 471.

Broad, C. D. (1923). *Scientific Thought*. New York: Harcourt, Brace and Co.

Brown, H. (2002). *Physical Relativity*. Oxford: Oxford University Press.

Brown, H. R., and Wallace, D. (2005). "Solving the measurement problem: de Broglie–Bohm loses out to Everett." *Foundations of Physics* **35**, 517–40.

Brush, S. (1976). *The Kind of Motion We Call Heat: History of the Kinetic Theory of Gases in the Nineteenth Century*. Amsterdam: Elsevier.

Bub, J. (1997). *Interpreting the Quantum World*. Cambridge: Cambridge University Press.

Bub, J., Clifton, R. and Monton, B. (1997). "The bare theory has no clothes." In R. Healey and G. Hellman (eds), *Minnesota Studies in Philosophy of Science XVII*. University of Minnesota.

Butterfield, J. (2011). "On time *chez* Dummett." Preprint: http://philsci-archive.pitt.edu/8848/1/ChezDummettJNB.pdf. [A shorter version is forthcoming in a special issue of *European Journal of Analytical Philosophy* in honor of Michael Dummett.]

Callender, C. (2002). "Thermodynamic asymmetry in time." In E. N. Zalta (ed.), *The Stanford Encyclopedia of Philosophy* (Spring 2002 edition), URL: <http://plato.stanford.edu/archives/win2001/entries/time-thermo/>

Callender, C. (2010). "Is time an illusion?" *Scientific American* **302**(6), 58–65.

Chiatti, L. (1995). "The path integral and transactional interpretation." *Foundations of Physics* **25**, 481–90.

Clifton, R. and Halvorsen, H. (2000). "Are Rindler quanta real? Inequivalent concepts in quantum field theory." Preprint: http://philsci-archive.pitt.edu/73/1/rindler.pdf

Clifton, R. and Monton, B. (1999). "Losing your marbles in wavefunction collapse theories." *British Journal of Philosophical Science* **50**(4), 697–717.

Cramer, J. G. (1980). "Generalized absorber theory and the Einstein–Podolsky–Rosen paradox." *Physical Review D* **22**, 362–76.

Cramer, J. G. (1983). "The arrow of electromagnetic time and the generalized absorber theory." *Foundations of Physics* **13**, 887–902.

Cramer, J. G. (1986). "The transactional interpretation of quantum mechanics." *Reviews of Modern Physics* **58**, 647–88.

Cramer, J. G. (1988). "An overview of the transactional interpretation." *International Journal of Theoretical Physics* **27**, 227.

Cramer, J. G. (2005). "The quantum handshake: a review of the transactional interpretation of quantum mechanics." Presented at "Time-Symmetry in Quantum Mechanics" Conference, Sydney, Australia, July 23, 2005. Available at: http://faculty.washington.edu/jcramer/PowerPoint/Sydney_20050723_a.ppt

Davies, P. C. W. (1970). "A quantum theory of Wheeler–Feynman electrodynamics." *Proceedings of the Cambridge Philosophical Society* **68**, 751.

Davies, P. C. W. (1971). "Extension of Wheeler–Feynman quantum theory to the relativistic domain I. Scattering processes." *Journal of Physics A: General Physics* **6**, 836.

Davies, P. C. W. (1972). "Extension of Wheeler–Feynman quantum theory to the relativistic domain II. Emission processes." *Journal of Physics A: General Physics* **5**, 1025–36.

de Broglie, L. (1923). *Comptes Rendues* **177**, 507–10.

de Broglie, L. (1924). "On the theory of quanta." PhD Dissertation.

Deutsch, D. (1999). "Quantum theory of probability and decisions." *Proceedings of the Royal Society of London* **A455**, 3129–37.

Devitt, M. (1991). *Realism and Truth*, 2nd edn. Oxford: Blackwell.

DeWitt, B. (1970). "Quantum mechanics and reality: could the solution to the dilemma of indeterminism be a universe in which all possible outcomes of an experiment actually occur?" *Physics Today* **23**, 30–40.

DeWitt, B. (2003). *The Global Approach to Quantum Field Theory*, Vol. **2**. Oxford: Oxford University Press.

Dirac, P. A. M. (1927). "The quantum theory of the emission and absorption of radiation," *Proceedings of the Royal Society, Series A* **114**, 243–65.

Dirac, P. A. M. (1938). *Proceedings of the Royal Society, Series A* **167**, 148–68.

Dolce, D. (2011). "de Broglie deterministic dice and emerging relativistic quantum mechanics." *Journal of Physics: Conference Series* **306**, 012049.

Dorato, M. and Felline, L. (2011). "Scientific explanation and scientific structuralism." In A. Bokulich and P. Bokulich (eds), *Scientific Structuralism*. Boston Studies in Philosophy of Science. Berlin: Springer. Preprint: http://philsciarchive.pitt.edu/5095/1/Structural_Explanationsfinal.pdf

Dowe, P. (2000). *Physical Causation*. New York: Cambridge University Press.

Dummett, M. (1964). "Bringing about the past." *Philosophical Review* **73**(3), 338–59.

Dyson, F. (2009). "Birds and frogs." AMS Einstein lecture. *Notices of the AMS* **56**, 212–23.

Earman, J. (1986). "Why space is not a substance (at least not to first degree)." *Pacific Philosophical Quarterly* **67**, 225–44.

Earman, J. (2008). "Reassessing the prospects for a growing block model of the universe." *International Studies in Philosophy of Science* **22**, 135–64.

Einstein, A. (2010). *Sidelights on Relativity*. Whitefish, MT: Kessinger Publishing.

Einstein, A., Podolsky, B. and Rosen, N. (1935). "Can quantum-mechanical description of physical reality be considered complete?" *Physical Review* **47**(10), 777–80.

Eisberg, R. and Resnick, R. (1974). *Quantum Physics of Atoms, Molecules, Solids, Nuclei, and Particles*. New York: John Wiley & Sons.

Elitzur, A. C., and Vaidman, L. (1993). "Quantum mechanical interaction-free measurements." *Foundations of Physics* **23**, 987–97.

Elitzur, A. C., Dolev, S. and Zeilinger, A. (2002). "Time-reversed EPR and the choice of histories in quantum mechanics." In Proceedings of XXII Solvay Conference in Physics, Special Issue, *Quantum Computers and Computing*, 452–61.

Englert, F. and Brout, R. (1964). "Broken symmetry and the mass of gauge vector mesons." *Physical Review Letters* **13**, 321–3.

Everett, H. (1957). "Relative state formulation of quantum mechanics." *Reviews of Modern Physics* **29**, 454–62.

Falkenburg, B. (2010). *Particle Metaphysics*. Berlin: Springer.

Feynman, R. P. (1985). *QED: The Strange Theory of Light and Matter*. Princeton, NJ: Princeton University Press.

Feynman, R. P. (1998). *Theory of Fundamental Processes*. Boulder, CO: Westview Press.

Feynman, R. P., and Hibbs, A. R. (1965). *Quantum Mechanics and Path Integrals*. New York: McGraw-Hill.

Feynman, R. P., Leighton, R. and Sands, M. (1964). *The Feynman Lectures on Physics*, Vol. 3. New York: Addison-Wesley.

Fields, C. (2011). "Classical system boundaries cannot be determined within quantum Darwinism." *Physics Essays* **24**, 518–22.

Forrest, P. (2004). "The real but dead past: a reply to Braddon–Mitchell." *Analysis* **64**, 358–62.

Friedman, M. (1986). *Foundations of Space–Time Theories*. Princeton, NJ: Princeton University Press.

Frigg, R. (2005). "Review of Kuhlman, Lyre and Wayne (2002)." *Philosophy of Science* **72**, 511–14.

Ghirardi, G. C., Rimini, A. and Weber, T. (1986). *Physical Review D* **34**, 470.

Gisin, N. (2010). "The free will theorem, stochastic quantum dynamics and true becoming in relativistic quantum physics," arXiv:1002.1392v1 [quant-ph].

Greaves, H. (2004) "Understanding Deutsch's probability in a deterministic multiverse." *Studies in History and Philosophy of Modern Physics* **35**, 423–56.

Grunbaum, A. (1973). *Philosophical Problems of Space and Time*. Dordrecht: Kluwer.

Guralnik, G. S., Hagen, C. R. and Kibble, T. W. B. (1964). "Global conservation laws and massless particles." *Physical Review Letters* **13**, 585–7.

Hacking, I. (1983). *Representing and Intervening: Introductory Topics in the Philosophy of Natural Science*. Cambridge: Cambridge University Press.

Hardy, L. (1992a). "Quantum mechanics, local realistic theories, and Lorentz-invariant realistic theories". *Physical Review Letters* **68** (20), 2981–4.

Hardy, L. (1992b). "On the existence of empty waves in quantum theory." *Physical Letters A* **167**, 11–16.

Heathwood, C. (2005). "The real price of the dead past: a reply to Forrest and to Braddon–Mitchell." *Analysis* **65**, 249–51.

Heisenberg, W. (1958). *Physics and Philosophy*. New York: Harper Row.

Heisenberg, W. (2007). *Physics and Philosophy*. Harper Perennial Modern Classics edition. New York: HarperCollins.

Hellwig, K. E., and Kraus, K. (1970). "Formal description of measurements in local quantum field theory." *Physical Review D* **1**, 566–71.

Higgs, P. W. (1964). "Broken symmetries and the masses of gauge bosons." *Physical Review Letters* **13**, 508–9.

Hoyle, F. and Narlikar, J. V. (1969). *Annals of Physics* **54**, 207–39.

Jammer, M. (1993). *Concepts of Space: the History of Theories of Space in Physics*. New York: Dover Books.

Kant, I. (1996). *Critique of Pure Reason*. Indianapolis: Hackett. English translation, Werner Pluha.

Kastner, R. E. (1999). "Time-symmetrized quantum theory, counterfactuals, and 'advanced action'." *Studies in History and Philosophy of Modern Physics* **30**, 237–59.

Kastner, R. E. (2006). "Cramer's transactional interpretation and causal loop problems." *Synthese* **150**, 1–14.

Kastner, R. E. (2011a). "de Broglie waves as the 'bridge of becoming' between quantum theory and relativity." *Foundations of Science*, forthcoming. doi:10.1007/s10699–011-9273-4.

Kastner, R. E. (2011b). "On delayed choice and contingent absorber experiments." *ISRN Mathematical Physics*, Vol. **2012**, Article ID 617291, 9 pages, 2012. doi:10.5402/2012/617291.

Kastner, R. E. (2011c). "The broken symmetry of time." *AIP Conference Proceedings*, Vol. **1408**, pp. 7–21. doi:10.1063/1.3663714.

Kent, A. (2010). "One world versus many: the inadequacy of Everettian accounts of evolution, probability, and scientific confirmation." In S. Saunders, J. Barrett, A. Kent and D. Wallace (eds), *Many Worlds? Everett, Quantum Theory and Reality*. Oxford: Oxford University Press.

Kim, Y-H., Yu, R., Kulik, S. P., Shih, Y. H. and Scully, M. (2000). "A delayed choice quantum eraser." *Physical Review Letters* **84**, 1–5.

Kondo, J. (1964). "Resistance minimum in dilute magnetic alloys." *Progress of Theoretical Physics* **32**, 37.

Konopinski, E. J. (1980). *Electromagnetic Fields and Relativistic Particles*. New York: McGraw-Hill.

Ladyman, J. (2009). "Structural realism." In E. N. Zalta (ed.), *The Stanford Encyclopedia of Philosophy* (Summer 2009 edition). URL: <http://plato.stanford.edu/archives/sum2009/entries/structural-realism/>
Lamb, W. and Retherford, R. (1947). "Fine structure of the hydrogen atom by a microwave method." *Physical Review* **72**(3), 241–3.
Landau, L. D., and Peierls, R. (1931). *Zeitschrift für Physik* **69**, 56.
Lewis, D. (1986). *On the Plurality of Worlds*. Oxford: Blackwell.
Lewis, P.J. (2007). "How Bohm's theory solves the measurement problem." *Philosophy of Science* **74**, 749–60.
MacKinnon, E. (2005). "Generating ontology: from quantum mechanics to quantum field theory." Preprint: http://philsci-archive.pitt.edu/2467/1/Ontology.pdf
Marchildon, L. (2006). "Causal loops and collapse in the transactional interpretation of quantum mechanics." *Physics Essays* **19**, 422.
Marchildon, L. (2008). "On relativistic element of reality." *Foundations of Physics* **38**, 804–17.
Maudlin, T. (1995), "Why Bohm's theory solves the measurement problem." *Philosophy of Science* **62**, 479–83.
Maudlin, T. (2002). *Quantum Nonlocality and Relativity: Metaphysical Intimations of Modern Physics*, 2nd edn. Oxford: Wiley-Blackwell.
McMullin, E. (1984). "A case for scientific realism." In J. Leplin (ed.), *Scientific Realism*. Berkeley, CA: University of California Press.
McTaggart, J. E. (1908). "The unreality of time." *Mind: A Quarterly Review of Psychology and Philosophy* **17**, 456–73.
Mermin, D. (1989). "What's wrong with this pillow?" *Physics Today* **42**(4), 9.
Miller, D. (2011). Private communication.
Neihardt, J. G. (1972). *Black Elk Speaks*. New York: Washington Square Press.
Norton, J. (2010). "Time really passes." *Humana Mente: Journal of Philosophical Studies* **13**, 23–34.
Orozco, L. (2002). "A double-slit quantum eraser experiment." http://grad.physics.sunysb.edu/~amarch/ [accessed September 6, 2011].
Pegg, D. T. (1975). *Reports on Progress in Physics* **38**, 1339.
Penrose, R. (1989). *The Emperor's New Mind: Concerning Computers, Minds, and Laws of Physics*. Oxford: Oxford University Press.
Petersen, A. (1963). "The philosophy of Niels Bohr." *Bulletin of the Atomic Scientists* **19**(7).
Price, H. (1996). *Time's Arrow and Archimedes' Point*. Oxford: Oxford University Press.
Psillos, S. (1999). *Scientific Realism: How Science Tracks Truth*. London: Routledge.
Psillos, S. (2003). *Causation and Explanation*. Montreal: McGill-Queens University Press.
Pusey, M., Barrett, J. and Rudolph, T. (2011). "The quantum state cannot be interpreted statistically." Preprint: quant-ph/1111.3328
Putnam, H. (1967). "Time and physical geometry." *Journal of Philosophy* **64**, 240–47. [Reprinted in Putnam's *Collected Papers*, Vol. I. Cambridge: Cambridge University Press, 1975.]
Quine, W. (1953). *From a Logical Point of View*. Boston, MA: Harvard University Press.
Redhead, M. (1995). "More ado about nothing." *Foundations of Physics* **25**, 123–37.
Reichenbach, H. (1958). *The Philosophy of Space and Time*. New York: Dover Books. Trans. M. Reichenbach and J. Freund.
Rietdijk, C. W. (1966). "A rigorous proof of determinism derived from the special theory of relativity." *Philosophy of Science* **33**, 341–4.
Rovelli, C. (1996). "Relational quantum mechanics." *International Journal of Theoretical Physics* **35**(8), 1637–78.

Russell, B. (1913). "On the notion of cause." *Proceedings of the Aristotelian Society* **13**, 1–26.

Russell, B. (1948). *Human Knowledge*. New York: Simon and Schuster.

Russell, B. (1959). *The Problems of Philosophy*. Oxford: Oxford University Press.

Sakurai, J. J. (1973). *Advanced Quantum Mechanics*, 4th edn. New York: Addison-Wesley.

Sakurai, J. J. (1984). *Modern Quantum Mechanics*. New York: Addison-Wesley.

Salmon, W. (1984). *Scientific Explanation and the Causal Structure of the World*. Princeton, NJ: Princeton University Press.

Salmon, W. (1997). "Causality and explanation: a reply to two critiques." *Philosophy of Science* **64**(3), 461–77.

Savitt, S. (2008). "Being and becoming in modern physics." In E. N. Zalta (ed.), *The Stanford Encyclopedia of Philosophy* (Winter 2008 edition). URL: <http://plato.stanford.edu/archives/win2008/entries/spacetime-bebecome/>

Schlosshauer, M. and Fine, A. (2003). "On Zurek's derivation of the Born Rule." *Foundations of Physics* **35**(2), 197–213.

Schrödinger, E. (1935). "The present situation in quantum mechanics." *Proceedings of the American Philosophical Society* **124**, 323–38.

Shimony, A. (2009). "Bell's theorem." *Stanford Encyclopedia of Philosophy*, http://plato.stanford.edu/entries/bell-theorem/#7

Silberstein, M., Stuckey, W. M. and Cifone, M. (2008). "Why quantum mechanics favors adynamical and acausal interpretations such as relational blockworld over backwardly causal and time-symmetric rivals." *Studies in the History and Philosophy of Modern Physics* **39**, 736–51.

Sklar, L. (1974). *Space, Time and Spacetime*. Berkeley, CA: University of California Press.

Spekkens, R. (2007). "Evidence for the epistemic view of quantum states: a toy theory." *Physical Review A* **75**, 032110.

Stapp, H. (2011) "Retrocausal effects as a consequence of orthodox quantum mechanics refined to accommodate the principle of sufficient reason." Preprint: http://arxiv.org/abs/1105.2053

Stein, H. (1968). "On Einstein–Minkowski space-time." *The Journal of Philosophy* **65**, 5–23.

Stein, H. (1991). "On relativity theory and openness of the future." *Philosophy of Science* **58**, 147–67.

Stewart, I. and Golubitsky, M. (1992). *Fearful Symmetry: Is God A Geometer?* Oxford: Blackwell.

Sutherland, R. (2008). "Causally symmetric Bohm model." *Studies in History and Philosophy of Modern Physics* **39**, 782–805.

Teller, P. (1997). *An Interpretive Introduction to Quantum Field Theory*. Princeton, NJ: Princeton University Press.

Teller, P. (2002). "So what is the quantum field?" In M. Kuhlman, H. Lyre and A. Wayne (eds), *Ontological Aspects of Quantum Field Theory*. Singapore: World Scientific Publishing, pp. 145–60.

Tooley, M. (1997). *Time, Tense and Causation*. Oxford: Clarendon Press.

Tumulka, R. (2006). "Collapse and relativity." In A. Bassi, D. Duerr, T. Weber and N. Zanghi (eds), *Quantum Mechanics: Are there Quantum Jumps?* AIP Conference Proceedings 844, American Institute of Physics, pp. 340–52. Preprint: http://arxiv.org/PS_cache/quant-ph/pdf/0602/0602208v2.pdf

Valentini, A. (1992). "On the pilot-wave theory of classical, quantum and subquantum physics." PhD Dissertation. Trieste: International School for Advanced Studies.

van Fraassen, B. (1991). *Quantum Mechanics: An Empiricist View.* Oxford: Oxford University Press.

van Fraassen, B. (2004). *The Empiricist Stance*. New Haven, CT: Yale University Press.

Wallace, D. (2006). "Epistemology quantized: circumstances in which we should come to believe in the Everett interpretation." *British Journal for the Philosophy of Science* **57**, 655–89.

Weingard, R. (1972). "Relativity and the reality of past and future events." *British Journal for the Philosophy of Science* **23**, 119–21.

Wheeler, J. A. and Feynman, R. P. (1945). "Interaction with the absorber as the mechanism of radiation." *Reviews of Modern Physics* **17**, 157–61.

Wheeler, J. A. and Feynman, R. P. (1949). "Classical electrodynamics in terms of direct interparticle action." *Reviews of Modern Physics* **21**, 425–33.

Wheeler, J. A. and Zurek, W. H. (1983). *Quantum Theory and Measurement.* Princeton Series in Physics. Princeton, NJ: Princeton University Press.

Wilson, K. G. (1971). "Feynman graph expansion for critical exponents." *Physical Review Letters* **28**, 548.

Wilson, K. G. (1974). *Physical Review D* **10**, 2445.

Wilson, K. G. (1975). "The renormalization group: critical phenomena and the Kondo problem." *Reviews of Modern Physics* **47**, 773–840.

Worrall, J. (1989). "Structural realism: the best of both worlds?" *Dialectica* **43**, 99–124.

Zee, A. (2010). *Quantum Field Theory in a Nutshell.* Princeton, NJ: Princeton University Press.

Zeh, H. D. (1989). *The Physical Basis of the Direction of Time*. Berlin: Springer-Verlag.

Zurek, W. H. (2003). "Decoherence, einselection, and the quantum origins of the classical." *Reviews of Modern Physics* **75**, 715–75.

Index